WIRELESS AND SATELLITE TELECOMMUNICATIONS:
THE TECHNOLOGY, THE MARKET & THE REGULATIONS

Feher/Prentice Hall Digital and Wireless Communication Series

Carne, E. Bryan. Telecommunications Primer: Signal, Building Blocks and Networks

Feher, Kamilo. Wireless Digital Communications: Modulation and Spread Spectrum Applications

Pelton, N. Joseph. Wireless Satellite Telecommunications: The Technology, the Market & the Regulations

Other Books by Dr. Kamilo Feher

Advanced Digital Communications: Systems and Signal Processing Techniques

Telecommunications Measurements, Analysis and Instrumentation

Digital Communications: Satellite/Earth Station Engineering

Digital Communications: Microwave Applications

Available from CRESTONE Engineering Books, c/o G. Breed, 5910 S. University Blvd., Bldg. C-18 #360, Littleton, CO 80121, Tel. 303-770-4709, Fax 303-721-1021, or from DIGCOM, Inc., Dr. Feher and Associates, 44685 Country Club Drive, El Macero, CA 95618, Tel. 916-753-0738, Fax 916-753-1788.

WIRELESS AND SATELLITE TELECOMMUNICATIONS:

THE TECHNOLOGY, THE MARKET & THE REGULATIONS

DR. JOSEPH N. PELTON

For book and bookstore information

http://www.prenhall.com

Prentice Hall P T R
Upper Saddle River, NJ 07458

Library of Congress Cataloging-in-Publication Data

Pelton, Joseph N.
    Wireless and satellite telecommunications: the technology, the market,
  and the regulations / by Joseph N. Pelton.
        p. cm. -- (Feher/Prentice Hall digital and wireless communication series
    Includes bibliographical references and index.
    ISBN 0-13-140493-8 (alk. paper)
      1. Wireless communication systems.   2. Artificial satellites in telecommunications.
  3. Telecommunications equipment industry--Government policy--United States.
  I. Title. II. Series.
  HE9719.P423   1995
  384.5--dc20
                                               95-7049
                                               CIP

Editorial/production supervision & design: *Patti Guerrieri*
Cover designer: *DeFranco Design, Inc.*
Manufacturing buyer: *Alexis R. Heydt*
Acquisitions editor: *Karen Gettman*
Editorial assistant: *Barbara Alfieri*

©1995 by Prentice Hall PTR
Prentice-Hall Inc.
A Simon & Schuster Company
Upper Saddle River, NJ 07458

The publisher offers discounts on this book when ordered in bulk quantities.
For more information, contact:

    Corporate Sales Department
    Prentice Hall PTR
    One Lake Street
    Upper Saddle River, NJ 07458
    Phone: 800-382-3419
    Fax: 201-236-7141
    e-mail: corpsales@prenhall.com

All products or services mentioned in this document are the trademarks or service marks of their respective companies or organizations.

Printed in the United States of America
10 9 8 7 6 5 4 3 2 1

ISBN   0-13-140493-8

Prentice-Hall International (UK) Limited, *London*
Prentice-Hall of Australia Pty. Limited, *Sydney*
Prentice-Hall of Canada Inc., *Toronto*
Prentice-Hall Hispanoamericana, S.A., *Mexico*
Prentice-Hall of India Private Limited, *New Delhi*
Prentice-Hall of Japan, Inc., *Tokyo*
Simon & Schuster Asia Pte. Ltd., *Singapore*
Editora Prentice-Hall do Brasil, Ltda., *Rio de Janeiro*

DEDICATION

*This book is dedicated to Eloise C. Pelton
my wife, my editor, and my friend.*

TABLE OF CONTENTS

PREFACE

The purpose of this book is to provide an in-depth, up-to-date, and comprehensive understanding of wireless and satellite telecommunications. This means a presentation on wireless technology, on emerging wireless markets, on key regulatory policies, as well as on services and applications in the field. Unlike most books in this field, the main focus is not on technology. The objective is, in fact, to present only that technology that is sufficient to define the market and regulatory aspects of the field of wireless and satellite telecommunications. This still means, however, exploring the basic technologies involved with satellites, licensed and unlicensed cellular radio, over-the-air radio and television broadcasting, specialized mobile radio (SMR) services, high-tier and low-tier personal communications service (including unlicensed PCS), wireless LANs and PABXs, microwave relay, terrestrial cellular television, and infrared bus wireless services. In this respect, this technological overview will explore wireless and satellite telecommunications as it operates today and as it is to function tomorrow. This book gives special attention to cellular telecommunications and satellites, but for the sake of economical word usage the general phrase "wireless" will be used throughout the book. An attempt will be made, however, to define the context and which special element of the very broad wireless and satellite field is indeed meant.

The key issues and concepts needed to plan, design, receive regulatory approval for, and actually implement wireless telecommunications systems will thus receive primary emphasis. More technical texts will thus address such issues as the optimum choice of modulation or multiplexing schemes, strategies for node interconnection of personal communications services, or coping with scintillation and multipath problems. This text does, however, provide a synoptic overview of all aspects of the field of wireless telecommunications, including markets, policy and regulation, standards, tariffs, and at least the basics of wireless and satellite technology. In particular it covers such aspects as emerging U.S. markets, current management issues, and contemporary American regulatory and policy frameworks. It is essentially designed for telecommunications courses where markets and regulatory policies are more important than the technology and especially for the businessman, attorney, or other non-engineers, who are just entering this complex and exciting field.

The goals of this book are thus to be both up-to-date concerning the latest developments in the field of wireless communications as well as to be very practically oriented. This means placing special emphasis on the licensing process, on frequency allocations, and on how a new start-up venture might be initiated in this field. This in turn means addressing how official filings are made for wireless systems, how petitions are made within rule-making procedures and what criteria are used to decide upon competitive applications. This will also include exploring the implications of the new frequency auctioning procedures in the United States and related frequency allocation and regulatory issues in other countries. The international implications of the "auctioning" of frequencies will likely become better known during the World Administration Radio Conference in 1995. Key national, regional, and international standards will likewise be explored although the primary focus will be on the United States. Virtually all existing and future wireless services will be addressed to some degree, both in terms of the market trends and future service development.

This book thus covers a very wide range of subjects from satellites to cellular and more generally from today's wireless LANs to future microcellular and picocellular services and so-called Universal Personal Telecommunications Services. Although the primary focus will be on the application of radio frequencies, there will also be some attention devoted to infrared and free-space optical communications. The fundamentals of all these wireless services and their related technologies will be examined along with how they are expected to grow in terms of services and markets. In general terms wireless, satellite, and mobile services are expected to grow to perhaps 20 to 25 percent of the total global telecommunications market in the early part of the twenty-first century. In short this overall field will become increasingly important and claim an ever growing market share for many years to come.

The emphasis will typically be on future developmental trends. Thus some of the more "obsolescent" or "passé" radio wave applications such as tropo-scatter, High Frequency (HF) "short wave" radio for point-to-point communications or push-to-talk dispatching equipment will be only briefly described. This is simply because these applications are rapidly disappearing as important telecommunications tools.

The basic philosophy of this book is to look at wireless telecommunications in a holistic and interdisciplinary way. It therefore examines the interactive relationship of technology, services, economics, tariffs, standards, and policy and regulation. It then examines how all of these factors relate to the overall integrated marketplace. For example, the complex problems associated with frequency allocations at the national and international level will be examined and analyzed in the context of the related economics, tariffing, and other financial issues as well as the key technical, operational, and health considerations that are also involved. Policy and regulatory issues involving the FCC and other U.S. governmental agencies, regional policy-making bodies, and the global telecommunications policy entities such as the International Telecommunication Union (ITU) and the International Standards

Organization (ISO) will be related to key new market trends and emerging new applications.

In a "conventional" text on telecommunications, the basic technical descriptions are presented in great detail. This includes considerable rigor in terms of formulas, charts, or precise technical definitions. These presentations are as a result both well defined and usually highly objective. Gain or multipath calculations turn out the same each time.

As noted above, the nontechnical or policy, economic, and management considerations are the key focuses in this book. Because of their subjective nature these areas require greater nuance and interpretation. It is often the "soft" aspects of telecommunications that tend to create problems of precision and consistency. This seems to be especially so in the field of wireless systems. The information presented about markets, services, standards, economics, policies, and regulations is not always neat, unambiguous, or even provable. This is because these nontechnical aspects are essentially very rapidly changing and often highly political. Differences in industrial interest, conflicting national objectives, clashes between and among standards-making or policy-making bodies, either within a country or in the international arena, can make outcomes unpredictable. In such an environment, outcomes can often be subject to rapid change. In other cases it can even be quixotic, haphazard, or internally inconsistent. In some cases regulatory processes can even be counterproductive to the development of new technology, new services, or even better or lower cost telecommunications.

The world, and certainly the world of wireless telecommunications, is not always fully rational. Public policy need not be always optimized against clear-cut algorithms. For these reasons the highly complex world of "telecommunications" when examined from an interdisciplinary perspective in terms of all of its components can sometimes be difficult to comprehend. When arcane and difficult to assess political strategies are at work, the key actor's motives or purposes may not be clear or direct. Several entities or groups may well be covertly interacting to maneuver a standard, a trade policy, or a tariffing concept in a different direction. They may take a particular position in an attempt to trigger a counterreaction. This seemingly "irrational" behavior may nevertheless stimulate a result that ultimately works to their longer-term or broader advantage. In short, things may not always be as they seem. Cause and effect may in such instances be hard to decipher. In the policy area, the way from A to C may not involve B at all. In such an environment nonlinear equations and chaos theory may well replace field theory or differential equations in predicting outcomes.

It is for these reasons that every effort will be taken to present as much factual information as possible throughout this text in terms of basic statistics, substantiated information, and known standards. Clearly quantifiable data and comparisons will be provided wherever possible. Further, special care will be used to cite sources, build objective foundations, and provide figures, charts, or numerical data whenever feasible. Also a glossary of terms is provided in Appendix A to provide definitions and background information for important concepts, acronyms, and phrases.

In cases where differences of viewpoint or perception apply, attempts will be made to present conflicting viewpoints. When particularly complex or subjective areas are addressed that cannot be fully examined in the text , then recommended appropriate further readings will be indicated wherever possible. In overall philosophy this book will seek to be as objective as possible and when the subjective and political elements are truly unavoidable the subjective points will be, as far as possible, placed into clear relief.

Finally, attempts will also be made to predict future market trends based upon reasonably clear data points or at least explicit assumptions. No one, of course, can consistently and accurately predict the future. At least, however, the basis of predictions can and will be clearly outlined. Just remember, when reviewing future predictions contained herein, the rather wise and wry observation of the noted futurist Arthur C. Clarke: "The future is not what it used to be." This ironic observation is more than a truism. It suggests among other things that history is certainly not always a prologue to the future and most predictions are not to be taken too seriously. Forecasts and predictions are only that. Only infrequently do they represent clear and accurate roadmaps to the future.

Please read the following text and enjoy where and when you can. Attempts to educate have, at least in some part, been tempered with the intent to build interest and once or twice even to amuse.

Joseph N. Pelton
Boulder, Colorado

PART I

UNDERSTANDING WIRELESS

CHAPTER ONE

INTRODUCTION

1.1 THE BASIC DYNAMICS OF TELECOMMUNICATIONS

The basic premise of this book is that telecommunications is much more than the study of technology. The many different elements that constitute the field of telecommunications interact in highly complex ways with technology being only a part of the process. The simple heuristic model presented below to explore the field of wireless telecommunications is used throughout this book to define the key components of the field. This model is thus based upon the so-called five key "drivers" of telecommunications that are seen as interacting with each other and upon a complex and changing global marketplace. These five drivers are: (a) technology; (b) services and applications; (c) standards; (d) economics, tariffs, and management systems; and (e) policy and regulation. The "marketplace," comprised of individuals, groups, and institutional and industrial consumers is influenced by each one of these drivers. These in turn also influence each other, both individually and collectively.

One could alter or amend this model to add other aspects such as quality control, globalism, or other relevant factors, but the model described in Figure 1.1 is believed to be both comprehensive and yet still focused enough to highlight the key elements needed to provide an in-depth understanding of the past, present, and future of telecommunications—especially of wireless telecommunications. This model is felt to be particularly relevant to wireless telecommunications because it is highly market-focused. Unlike conventional telecommunications where governmentally-mandated and directed investment and tariffs are predominant, wireless telecommunications represent a much more consumer-dependent and free-market-dependent enterprise.[1]

One could attempt to develop a model of telecommunications unique to wireless service, but this would be in some sense futile. This is because the information environment

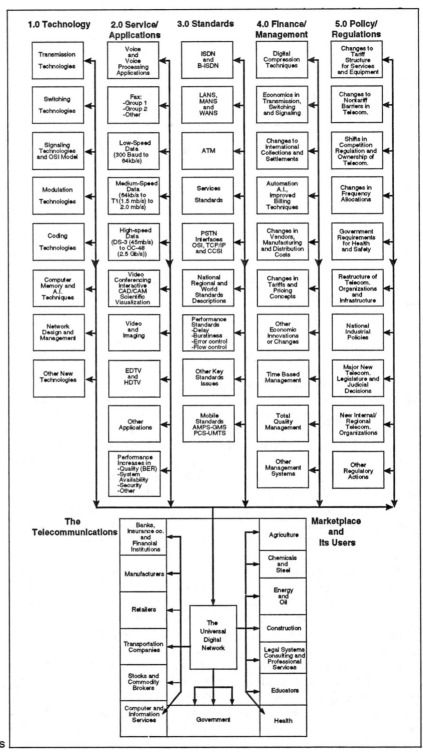

FIGURE 1.1
The Five Key "Drivers" Model of Telecommunications

of the twenty-first century seems very likely to be increasingly "hybrid." This is to say that virtually every message, call, or transmission on future networks will be routed through links that are part wire and part wireless—sometimes changing back and forth several times. This, as will be explored later, is what the new future-oriented concept known as Universal Personal Telecommunications is all about. This detailed yet nevertheless simplified model of the field of communications as given in Figure 1.1 will be used throughout this book to help examine how all the forces in the field of telecommunications interact with one another. This is believed to be especially true in terms of wireless network development, financing, implementation, and use. The general applicability of this model to all types of telecommunications—both wire and wireless—is thus seen as more of an advantage rather than as a limitation. This is because the future of telecommunications seems likely to represent a merger of the two basic technologies into complex interlinked networks. If there is one message in this book, it is that hybrid networks will be increasingly important and pervasive. This trend is projected on the basis of the need to combine the broad band throughput and performance of fiber optics with the flexibility, mobility, and accessibility of wireless.

1.2 WIRELESS TELECOMMUNICATIONS: WHAT IS IT?

Wireless telecommunications is the process of communicating information over a distance through the free-space environment rather than by using a wire or other physical conduit and doing so by the use of electromagnetic signals. This covers many diverse possibilities. It means the use of modulated radio wave frequencies to broadcast signals such as for television or radio. It can likewise mean radiated transmissions between and among antennas for two-way or limited-party interactive communications. Or it can also mean the use of sound waves such as sonar, infrared signals for hand-held television remote controls or even modulated laser signals for intersatellite relays in space. Optical or laser communications does not necessarily mean fiber optic communications.[2]

The distinguishing feature of wireless telecommunications as opposed to wire communications is that the signal is radiated into a free-space environment rather than confined to a wire, a coaxial cable, a fiber optic link, or even tin cans and a taut string. The advantage of this is that anyone can receive the signal in the broadcast area whether they are at a fixed location or in a mobile unit. This also allows new links to be established at a moment's notice since physical links do not have to be installed. The disadvantage is that the "frequency" is used up for the entire transmitting zone in contrast to a wire transmission that restricts the signal to inside the conduit. It also means limits on privacy of communications, on priority of access, and on exclusivity of use.

Wire thus "conserves" frequency while wireless does not. It is for this reason that so-called cellular transmission transmissions have become important. This is because the signal is low in power or flux density and confined to a small zone. Thus, the use or consumption of the assigned frequency is geographically limited to a small area. This creation of lots of small zones for frequency use in wireless applications allows high levels of frequency reuse by recycling the frequencies used within the various small zones or cells. This in turn effectively boosts the capacity and effective band-width of wireless systems. The concept of frequency reuse will be examined later since it is critical to so many different systems including cellular, personal communications services, satellites, and several other applications.

It is important to note at this point only that intensive reuse concepts are critical to the future growth and expansion of wireless service in the twenty-first century. In fact, the two most important factors with regard to the future development of wireless telecommunications are likely to be hyperintensive frequency reuse techniques and advanced digital compression techniques. In the field of satellite communications new and innovative reuse techniques may prove critical to the future ability to deploy large numbers of geosynchronous, medium earth orbit and low earth orbit satellites needed to respond to new market demands.

1.3 THE TELECOMMUNICATIONS MARKET AND NEW SERVICES

The two primary keys to the Five Drivers Model of telecommunications are to be found in the market and the new services and applications. These components, because they define "why" we want and need telecommunications, are key. In fact, some would argue that these are also the most crucial issues. Even if these were not the most critical, the market and service sectors still seem a very logical place to start the investigation of wireless telecommunications. In a competitive environment, it is often very wise to start by considering what the customer wants.

The meteoric rise in wireless telecommunications services in recent years was largely unanticipated by virtually all market forecasters. Especially, the spectacular rise of cellular telephone service starting in the 1980s was "underpredicted" by analysts, government officials, equipment suppliers, and service providers alike. The sustained market expansion rates for cellular telephone service have exceeded 25 percent per annum since 1984—sometimes by sizable margins. This suggests an almost pure example of a customer-demand-driven market. In the last few decades there have been few counterpart markets that have grown so rapidly and in response to customers "wanting and needing" the service. It has been far more common for the consumer to be "sold" a service that the telecommunications carrier or service provider "thought" the customer might need. The other comparable break-away services representing exponential growth and nearly unrestrained customer enthusiasm have been in the areas of facsimile (or fax), of private networks in the form of local area networks and wide area networks, and of cable television entertainment systems. The new direct broadcast satellite service also known as small dish television gives promise of becoming yet another such phenomena, although this trend is far from certain.[3]

It is important to note that none of the four exponential growth services listed above were based on any truly new technology. In fact, the evolution of the technologies associated with these four service areas—fax, LANs and WANs, cable television, and cellular telephone—has come **after** spectacular market success was achieved—not before. The following graphs (Figures 1.2a, 1.2b, 1.2c, and 1.2d) describe the market growth for each service plus indications of why they have been so successful in the market.[4], [5], [6], [7]

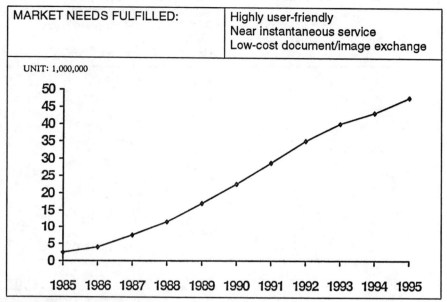

FIGURE 1.2A Market Success of Facsimile

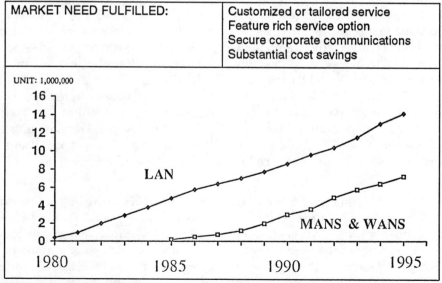

FIGURE 1.2B Market Success of LANs, MANs, and WANs

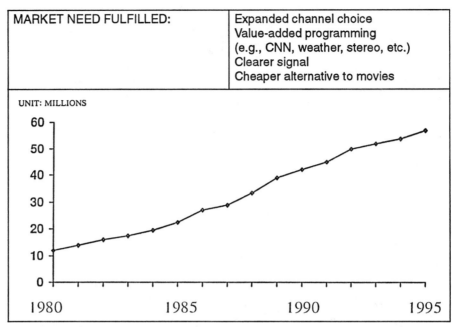

MARKET NEED FULFILLED:	Expanded channel choice Value-added programming (e.g., CNN, weather, stereo, etc.) Clearer signal Cheaper alternative to movies

UNIT: MILLIONS

FIGURE 1.2C   Market Success of Cable Television

The argument as to whether technological innovation or customer demand is the key to new market developments is in many ways a futile debate. Often it is a combination of "market pull" and "technology push" that create a new service or application. Usually, however, there has to be some real market interest to create a major new breakthrough. In some cases the first phase of a new service is heavily driven by new market demand such as in the case of cellular radio telephone service for mobile business people—the so-called "road warriors." In the second phase, which might be illustrated in this case by the recent emergence of Cellular Digital Packet Data (CDPD), digital cellular and soon personal communications services, the role of market demand and technology can reverse. This is particularly so as the general consumer market becomes predominant and as "sales push" emerges as a more important market factor. In point of fact, general consumer use of cellular telephone service achieved crossover in 1993 to exceed business-related usage. It could be argued that sales initiative in the cellular field took over from market demand as of this date.

There is certainly some evidence that this second stage of growth of a new service tends to be much more heavily impacted by "technology and sales push." Five to ten years after the new product or service has achieved its initial successful market penetration, the sales initiative can emerge as the predominant market factor. In such an environment the consumer is often presented with upgrades, refinements, and extra "bells and whistles" as incentives to move from a basic product or service to a "new and improved" offering. This new option is inevitably considered to be of higher value and certainly attract a higher price so as to differentiate the market.

MARKET NEED FULFILLED:	Anytime, anywhere personal communications Virtual office in a vehicle Emergency communications Status Flexibility and reconfigurability

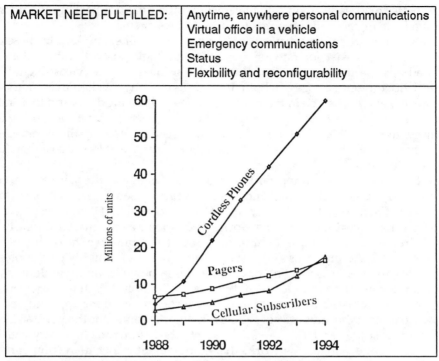

FIGURE 1.2D Market Success of Wireless Telephone Service

In today's technically advanced society, a new market demand that is truly perceived as essential and desirable by a broad segment of the business and consumer population will almost invariably attract significant financial resources. It is thus usually not hard to find second stage capital needed to develop new technology critical to bringing that initial product or service to the next level of sophistication once basic consumer and business demand is established. Initially, a new product or service in high demand may use relatively crude or unsophisticated technology such as the cases of the initial Group 1,2, and 3 fax or early "mom and pop" community antennas to support the 1950s version of cable television.[8]

Nothing encourages new investment into the required R&D for second stage development better than rapid market growth and eager consumer demand. Investors see such investments for product enhancements within a "hot" new market as a virtually guaranteed proposition with very low risk. In contrast, the lack of a clear perception of new market demand often slows new R&D efforts. The lack of a favorable consumer appreciation of the advantages of a new service like Direct Broadcast Satellite Television at least during its dormant period between 1975 and 1992 is a case in point. In this case the lack of consumer perception that DBS is somehow better, more user-friendly or cheaper than cable television has served as a deterrent to capital investment in this new service. As it now appears the new Hughes and Hubbard DBS systems give signals in late 1994 and early 1995 of becoming hugely successful. If indeed their marketing creates a broad consumer belief that DBS is somehow better or cheaper than cable television, then financial backing would suddenly become available in generous amounts for additional systems.[9]

As noted earlier, there are a number of "reverse" parallel examples . In these cases the services are seen, at least for now, as being "technology pushed" rather than "market pulled" Perhaps the most obvious of these are the Integrated Services Digital Network (ISDN) and videophone services. In both these cases the long awaited mass market has yet to materialize. All the research in the world to develop these products and their related applications are going to be largely for naught if the consumer demand concerning these services simply does not exist. Certainly, if other options for obtaining these services such as enterprise networks exist in terms of providing better service at lower cost then ISDN and video will lag behind.[10]

One of the reasons this is important to understand is that many wireless services, and cellular telephone and paging, in particular, seem to represent a classic example of a new yet basic market demand. It cannot be considered accidental that consumers have strongly embraced this new service around the world. The fundamental "market-friendly" characteristic of wireless service in order of importance are: (i) mobility; (ii) flexibility of establishing new remote interconnections; (iii) cost efficiency (at least for highly paid professionals); and (iv) ability to offer broadcasting or complex mesh networking capabilities as needed. In contrast, there are tremendous throughput capabilities and high-quality trenching advantages for fiber optic cable services for many telecommunications applications. In fact, it is generally true that the strengths of wireless networks represent the very weaknesses of wire or fiber technology and vice versa. The key to the future thus may very well be to design hybrid wire and wireless networks that complement rather than compete with each other.

1.4 FUTURE PATTERNS OF TELECOMMUNICATIONS GROWTH AND FREQUENCY USE

As much as cellular radio telephone service can be cited as an example of a consumer-driven new market, it must be recognized that there are other key factors in the growth and development of this service. These other ingredients are new standards, the new frequency reuse concepts inherent in cellular technology, digital compression, new frequency allocations, and other regulatory actions that enabled the offering of these services in an increasingly competitive and open market environment. Without these actions cellular could well have remained just a concept with limited practical applications.

Today, it is the competing claims for frequencies, in the Very High Frequency (VHF) (30–300 MHz) and especially in the Ultra High Frequency (UHF) (300–3000 MHz) range, that will determine much of the future growth of the wireless market over the next decade. The Super High Frequency (SHF) band from 3 to 30 GHz will, however, over time become increasingly important as seen in the strong interest in the 28-GHz band (i.e., 27.5 to 29.5 GHz) for cellular television as well as for satellite communications. There are a lot of conflicting vested interests as well as new aspirants for the use of all of these frequencies and especially for the VHF and UHF bands. The list of "contenders" include traditional over-the-air radio and television broadcasters, the mobile trucking and dispatching fleets plus taxi and limousine operators, the providers of aeronautical and maritime communications, instructional fixed television service providers, amateur radio, citizen's band services, walkie-talkie links, military communications, cellular mobile radio telephone service providers, prospective providers of personal communications networks, and mobile satellite operators.

These prospective users are for the most part in direct competition with one another, if not in terms of services at least in terms of frequency needs. Although the list of those in competition with one another is long and growing, the biggest fights today tend to focus on the bands from about 700 to 2500 MHz and largely involve the providers of mobile and broadcast services as well as terrestrial versus space-based service providers. Getting an exclusive license in a suitable band to provide a wireless telecommunications service is one of the largest challenges to operating a successful business in this field. New and more effective ways to share available bands will be critical to future commercial success in many mobile wireless and satellite services. The recipe for success in this field is not an easy one. It usually takes time, good technical advice and planning, proven financial capability, a previous successful corporate track record, extensive legal fees, patient dealings with state and U.S. governmental officials, and often liaison with international officials as well. The barriers to entry are not huge but they are not modest either.[11]

In short, not all wireless telecommunications service providers and operators are uniformly grouped together in a seamless and well-organized web of cooperation. It is actually quite the reverse. In fact, wireless communications services, with the exception of infrared and optical free-space communications, all require radio frequency allocations (i.e., both national and international) and most of these are in contention with one another.

There is an exception to this rule in that there are some limited frequency bands for socalled unlicensed wireless service. The so-called Part 15 bands are set aside by the FCC for commercial wireless use. These unlicensed bands are allocated on the basis of standards that keep the power sufficiently low so as to allow the use specific products or equipment within a particular office building or industrial site so as to not interfere with others using the same frequencies even on a nearby site. In-building Part 15 wireless communications in the form of wireless PABX services, unlicensed PCS frequencies, and even use of infrared frequencies for computer modem transmissions or remote controls for televisions are examples of unlicensed frequencies. In this respect, operators of wireless communication may also well be in conflict with other fixed communications, military operations, and scientific or other messaging services for frequency allocations. It is perhaps even more likely, however, that they may be in conflict with other wireless service providers.

The way future frequency allocations occur will likely vary a great deal on a country-by-country basis over the next few decades. Nevertheless these reallocations will indeed spread across the world and frequency auctions will be a clear part of the pattern. The U.S. practices will be closely scrutinized as the need and popularity of frequency allocations grows. The FCC has undertaken, in response to the Omnibus Budget Reconciliation Act (OBRA) of August 1993, a process for auctioning off a major block of frequencies for both narrow band and wide band Personal Communications Service. This new allocation of 200 MHz includes 50 MHz, which is currently being reallocated and another 150 MHz to follow over the next 15 years. These new frequencies reallocations will also cover both the newly defined Commercial Mobile Radio Services (CMRS) and the Private Mobile Radio Services

(PMRS). This provision of new PCS frequencies will likely allow a shift to the new digital PCS services and to begin a longer-term shift away from the currently pervasive analog systems. Existing digital cellular systems based on the EIA and TIA IS-55 standard can be expected to continue to grow. Space-based systems (e.g., Iridium, AMSC, Odyssey, Globalstar, Teledesic, Orbcom, INMARSAT, INTELSAT, CELLSAT, etc.) plus the new High Altitude Long Endurance (HALE) systems will seek PCS allocations and other new space-based frequency allocation for mobile and fixed services as well.

At this time what is clear is that the FCC will auction off the following frequencies in the phase one broad band PCS program. These are 2390–2400 MHz, 2402–2417 MHz; and 4660–4685 MHz. This provides an additional 25 MHz in the 2-GHz region and 25 MHz in the 4-GHz region. By February 1995, the federal government (i.e., the NTIA and the FCC) must identify the remaining 150 MHz of frequencies to be reallocated and auctioned off with half of these frequencies being in the 2-GHz band and half being above. In addition to these frequencies both the CELLSAT system and the American Mobile Satellite Corporation's subsidiary PCSAT are seeking allocations in the bands of 1970–1990 MHz and 2160–2180 MHz to provide PCS or supplemental PCS service. Although these changes will take a number of years to implement and will also involve the relocation of microwave systems, there is no doubt that a pattern of continuing and pervasive change in the frequency use will occur.[12]

This action by the U.S. government to reallocate frequencies to PCS, however, seems to have as much or more to do with obtaining billions of dollars in auction fees for the public coffers as with accelerating the move to new digital personal communications services and greater spectrum efficiency. Certainly it is odd that key regulatory policy for the field of telecommunications was accomplished through a budget act rather than a new communications law.[13]

In short, the shift of frequencies from wire to wireless applications and the reverse process of wireless to wire may well be accelerated by new market values realized through competitive bidding processes. At the macrolevel, Nicholas Negroponte of MIT Media Laboratories has suggested that shifts from older and lower-value services to new and higher-value services will occur rapidly over the next 15 years in what he characterizes as the "Negroponte Flip." In particular, this shift as shown in Figure 1.3 predicts the shift of radio frequencies from over-the-air broadcasting and satellite services to terrestrial wireless mobile telephone services. It also predicts the increased use of cable television for entertainment distribution and the corresponding decreased use of wire telephone services to the home and business.[14]

It seems clear that a shift in frequency allocations and in services, both on wire and wireless media, will occur. Historical and political patterns suggest, however, that the shifts may well not be as rapid or as sweeping as those suggested by the Negroponte Flip. An alternative view of these changes is presented in Figure 1.4 in terms of the so-called "Pelton Merge."[15]

This alternative view as presented in Figure 1.4 also anticipates the shift in services and delivery media, but on a more gradual basis and with a more intensive merging of the old and new. The importance of embedded plant, already deployed technology, and vested holdings in current frequency assignments is thus reflected in the Pelton Merge. In either

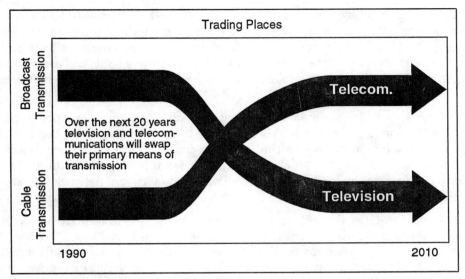

FIGURE 1.3 The "Negroponte Flip"

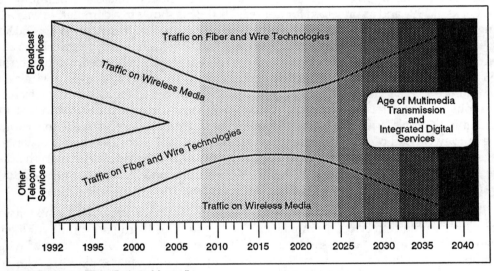

FIGURE 1.4 The "Pelton Merge"

scenario, however, significant shifts in how services are delivered in both the wire and wireless mode are still anticipated in the early part of the twenty-first century.

1.5 WIRELESS SERVICES AND APPLICATIONS

The breadth and depth of the world of wireless telecommunications is today not always well understood in a large part because the technology and services have grown and matured so rapidly in recent years. Furthermore the range of wireless services and applications is today constantly expanding. Consumers often do not recognize that products and services such as wireless remote controls for television, walkie-talkies, cellular telephone, radio, direct broadcast satellite services, television, microwave relay, and aeronautical and ship communications are all essentially within the same broad family of wireless technology.

All of the above, despite their dissimilarities, have some characteristics in common. They all, for instance, do share a host of important technical, operational, business, policy, and health considerations. These include concern for frequency allocations, health radiation standards, interference, rain attenuation, heat scintillation, multipath and ghosting problems, throughput, reliability, system availability, quality, and network operation, and a variety of interconnection issues.

The range of services available using wireless technology is amazingly broad. In terms of distance it can range from interplanetary communications to an "intraoffice bus" to provide service to a single small room. Wireless service might be a global satellite broadcast to billions of people or it may only involve mobile radio telephone service between a person in a parking lot to his nearby office. Figure 1.5 provides a composite summary presentation of more than a dozen component sectors of the overall annual wireless telecommunications market as of 1994 and as projected for the year 2000. This shows in a cumulative way a huge range of wireless service sectors that by the next century will represent an over $200 billion a year industry in terms of projected revenues. This would thus likely total about 20 percent of the overall worldwide telecommunications enterprise that is estimated to be about a $1 trillion per year industry at the turn of the century. This projected level of wireless telecommunications services is roughly equivalent to about half of today's global telecommunications market. These projections, which were based upon a survey of industry and academic experts carried out by the author are, in fact, more conservative than a comparative survey carried out by the International Engineering Consortium and the Center for Telecommunications Management at the University of Southern California. In the case of the IEC/CTM survey, growth rates for wireless cellular and PCS were projected to gain over a 25 percent share of the total telecommunications market by 2005.[17]

The prospects for wireless systems acting as both a counterbalance and an extension to fiber optic trunking services seem particularly bright. This is true now and will likely be even more true in the twenty-first century. The market strength of all forms of wireless technology as noted earlier are mobility, flexibility, cost efficiency, and networking interconnectivity. In general a doubling, or more, of wireless products and services seems likely in virtually all areas. The main exceptions are in the areas of over-the-air radio and television broadcasting and microwave relay services. These sectors will remain relatively static or may even decline with regard the general rate of inflation. It should be noted that the growth projected in Figure 1.5 for the microwave relay and distribution sector is essentially all in the specialized microwave distribution service (or wireless cable) and not in long-distance microwave relay. The only major concerns that readily stand out today concerning future market development of wireless technology and services are those six noted below:[18]

Wireless Market	Current Size of Market ($US)	Size of Market Year 2000 ($US)
Paging	<$1 Billion	>$1 Billion
Position Determination/Navigation	<$1 Billion	$2 Billion
Amateur Radio	<$1 Billion	>$1 Billion
Citizen Band /Walkie-Talkies, etc.	<$1 Billion	>$1 Billion
Cellular Radio Telephone	$35 Billion	$65 Billion
Personal Communications (PCS, UMTS, etc.)	<$ 1 Billion	$15 Billion
Specialized Mobile Radio	$ 3 Billion	$ 7 Billion
Radio and Television (Transmission Networks)	$20 Billion	$26 Billion
Wireless PBX /LANS	$ 2 Billion	$ 8 Billion
Fixed Satellite	$14 Billion	$23 Billion
Mobile Satellite	$2.5 Billion	$10 Billion
Direct Broadcast Satellite	$1.5 Billion	$ 6 Billion
Aeronautical/Maritime (Nonsatellite)	$ 2 Billion	$ 3 Billion
Military Mobile Communications (Terrestrial)	$10 Billion	$15 Billion
Military Mobile Communications (Satellite)	$ 7 Billion	$12 Billion
Microwave Relay and Distribution Systems	$3.5 Billion	$ 5 Billion
Interplanetary Communications	<$1 Billion	>$1 Billion
Wireless Emergency/Police	$6 Billion	$11 Billion
Industrial Wireless Products (e.g., Remote Controls, etc.)	<$1 Billion	>$1 Billion
TOTAL MARKET SIZE	$113.5 Billion	$210 Billion

FIGURE 1.5   Global Survey-Wireless Products and Services—1994–2000[16]

(a)   Health hazards from radio transmissions near and around humans.

(b)   Latency or transmission delay in satellite communications.

(c)   The increasing appetite for spectrum and frequency allocations for new and broad band radio and wireless telecommunications services.

(d)   The rapid improvement in the cost efficiency, transmission rates, and overall performance of potentially competitive fiber optic cable systems.

(e)   The increasing obsolescence of certain wireless technologies such as radio and television broadcasting, microwave relay, and so on.

(f)   The lack of universal standards for wireless communications and for seamless wireless to wire–line interconnection.

None of these problems appear to pose fundamental barriers to the further development of wireless technology. In fact, new capabilities such as infrared bus technology, low earth orbit satellites in global constellations, and new approaches to digital compression will likely spur intensive frequency reuse, better wireless node interconnection, increased performance, and possibly even enhance a competitive regulatory environment. This suggests that wireless technology could well continue its rapid growth well into the future. It also strongly suggests the possibility that hybrid wire and wireless systems may represent a key direction for the future.[19]

As important as markets and services clearly are, the issues of new technology as well as new policies and regulations must also be given careful consideration.

1.6 WIRELESS TECHNOLOGY

The basic concept of wireless telecommunication is almost deceptively easy to understand. An electromagnetic signal is created, modulated, amplified, and broadcast to one or more receivers that can be fixed or mobile. The basic issues that one must address in the design of wireless systems is common to all of telecommunications, namely the effective use of the available frequency spectrum and power to provide high-quality communications. Since wireless systems often involve mobile services, this implies a constantly changing environment with rapidly changing interference conditions and dynamically variable multipath reflections. This condition, plus the potential of conflicting demands for the use of radio frequencies in a free-space medium, means difficult challenges for creating high-quality signals.

The current uses of radio frequencies for wireless telecommunications include conventional fixed and mobile satellite communications, direct broadcast satellite services, scientific data relay from spacecraft, high frequency to millimeter wave-guide terrestrial radio relay, conventional radio and television broadcasting services, wireless cable television entertainment systems, instructional fixed television services, amateur radio services, specialized mobile radio services, analog and digital cellular radio telecommunications services, paging services, unlicensed band ISM services, wireless PABX and LAN services, emergency vehicle communications, disaster warning and relief systems, and military communications (terrestrial and space communications). Sometimes these services have separate frequency allocations from each other and sometimes they compete with each other or with other public telecommunications, industrial or scientific applications.

The rapid growth of wireless services has thus had the affect of creating rapidly growing demands on a limited resource. Furthermore, the desire to use frequencies in the Ultra High Frequency Band (UHF) (i.e., below 3 GHz) has served to compound this problem. The physical characteristics of the radio spectra in terms of "effective" wavelengths have tended to concentrate demand for most wireless services between the High Frequency (HF) bands and the Ultra High Frequency (UHF) band, although the next two bands in the Super High Frequency (SHF) and Extremely High Frequency bands are today starting to be used for wireless services.[18] Those radio frequency bands that are most commonly used for wireless communications today are thus as follows.

KEY FREQUENCY BANDS FOR WIRELESS COMMUNICATIONS

- High Frequency (HF)—3–30 MHz   (PRIME BAND)
- Very High Frequency (VHF)—30–300 MHz   (PRIME BAND)
- Ultra High Frequency (UHF)—300–3000 MHz   (PRIME BAND)
- Super High Frequency (SHF)—3–30 GHz   (INCREASING USE)
  (also known as microwave)
- Extremely High Frequency (EHF)—30–300 GHz   (PROSPECTIVE USE)
  (also known as millimeter wave)

Those frequencies above 3000 MHz are most typically used for satellite communications and terrestrial microwave even though so-called cellular television or LMDS in the millimeter wave are now planned for commercial services within the next few years. Indeed research and development to use much more intensively the millimeter frequencies for such purposes as advanced satellite communications and digital cellular broadcasting is actively underway. These higher frequencies are, however, subject to precipitation attenuation effects, atmospheric heat scintillations, and propagation distortions of other types. They are also unforgiving of any interruptions in direct line-of-sight connections in the transmission path. Frequencies below the high frequency (HF) range are, on the other hand, very limited in band-width. They thus offer limited utility for many future services, especially broad band services like television or high definition television. Even with highly innovative frequency reuse concepts and digital compression it seems unlikely that broad band applications for the VHF band or below would develop in the future simply because of the very limited spectra available.[20]

These conditions have combined to create a major problem in obtaining sufficient frequency allocations that are interference free or, at least, are at acceptable interference levels. The answer to the dilemma of limited available frequency bands has been severalfold. Most significantly, methods have been developed to allow intensive reuse of available frequency bands. The most common ways have been creation of geographically defined beams that are isolated from other beams. In satellite communications there are larger regional beams, zonal and very narrow, and very powerful spot beam antennas that allow for increasing degrees of frequency reuse. In terrestrial cellular radio telecommunications each "cell" is a geographically restricted coverage area. In the future with personal communications services there will be microcells or possibly even "picocells" that create very small coverage areas for extremely high levels of frequency reuse. The patterns of cell coverage for three types of cells are shown in the text that follows. The first is that of conventional cellular radio systems. The second is that for relatively low-powered microcellular systems. The third is for cellular satellite coverage from a low earth orbiting system. These are shown in Figures 1.6, 1.7, and 1.8, respectively. In each case it should be noted that these are idealized or theoretical diagrams rather than actually measured system performance.

Another important technique used to achieve frequency reuse is that of polarization discrimination. In this case the wanted frequency transmission is separated from the unwanted transmission by "polarizing" the signal. This concept is essentially the same as that used in polarized sunglasses. The polarization technique most often used is that of linear or orthogonal separation. That is to say the "wanted" and the "unwanted" transmission patterns are sent exactly at right angles to one another. Alternatively, there is circular polarization with one signal being sent with left-hand circularization and the other being

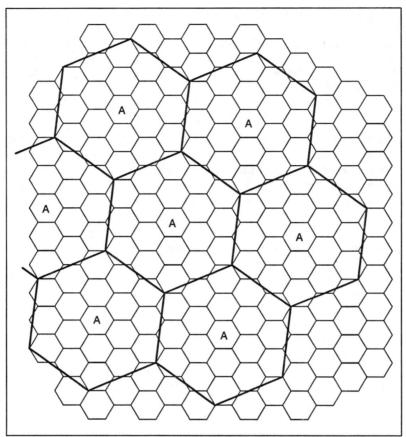

FIGURE 1.6   Cellular Coverage for Conventional Mobile Radio Service[21]

sent with right-hand polarization. It is possible with these combined techniques (i.e., spatial or cellular beam separation and polarization discrimination) to achieve dramatically increased reuse and thereby much greater capacities. Today there are plans for tenfold to even fortyfold reuse in some of the low earth orbit satellite systems. With advanced PCS systems, frequency reuse techniques that achieve one hundredfold, and ultimately even thousandfold reuse levels may be achievable. This can allow much more extensive, versatile, and broader band wireless applications including all forms of video and imaging services.[24]

There are other technologies that are also important to making more effective use of the limited frequency spectra that can be practically used in a free-space wireless environment. These involve the modulation, encoding, and multiplexing of signals. The detailed discussion of the various approaches such as Frequency Division Multiple Access (FDMA), Time Division Multiple Access (TDMA) and Code Division Multiple Access (CDMA) will be discussed later in this book. Also the use of the Phase Shift Keying (PSK) approach to create digitally encoded signals will be explained in Chapter 4.

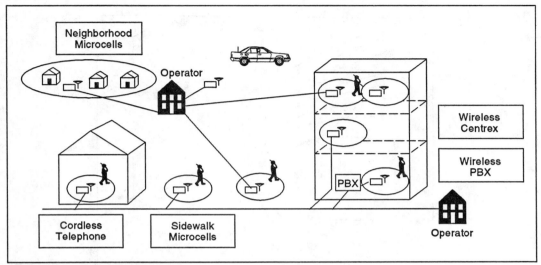

FIGURE 1.7 PCS Microcells[22]

At this stage it is important only to note that new modulation, coding, and multiplexing techniques can produce dramatically more reliable and higher-quality wireless transmission systems. The use of digital techniques, in particular, can especially help to create higher-capacity systems. First, this is because digital systems such as TDMA and CDMA are more resilient against interference and noise.[25] Second, digital compression techniques also allow more effective use of the available channels or frequency bands. In the area of voice services, digital compression techniques can be utilized to achieve acceptable service at bit rates as low as 4.8 kilobits per second. This is an improvement in efficiency of some ten times over conventional voice services.[26]

High-efficiency analog systems will, however, continue to be used in the 1990s. In particular, Single Side Band (SSB) techniques can yield results that are more efficient than conventional FDMA systems. In general, however, digital techniques and especially CDMA or spread spectrum systems will expand as this modulation matures and catches up to the technologically more mature TDMA systems. In short, both TDMA and CDMA will expand in general use and broad customer acceptance for terrestrial mobile services over the next decade.

TDMA will continue to be used in satellite systems, although CDMA systems will also be increasingly used here too. This may, however, tend to create problems after the year 2000 with regard to interfacing with terrestrial fiber optic systems. This is because optical systems are now tending toward Dense Wave Division Multiplexing (WDM) schemes or so-called "color multiplexing" techniques. These two multiplexing concepts are basically not very compatible with one another.[27]

The key element of telecommunications involves three key components: (a) transmission of information; (b) the switching of that information, if routed through an interactive and selectively connected network; and (c) signaling and network management systems to control intelligently the routing and processing of information within these networks.

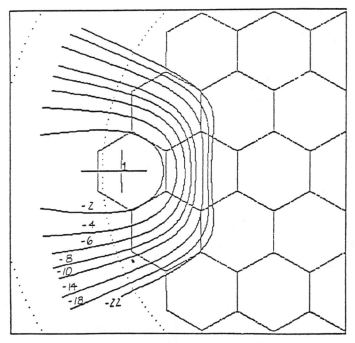

CELL 1 PATTERN: Motorola
Iridium

FIGURE 1.8   Cellular Coverage for a Low Earth Orbit Mobile
Satellite System[23]

When broadcasting information from a single point of origin to all receivers such as in the case of conventional radio and television stations or a traditional cable television system, the requirement is essentially only for transmission. The functions of signaling and switching are not particularly required. Today, however, almost all telecommunications systems are becoming more interactive. Integrated service networks that combine voice, data, e-mail, fax, radio, video, and imaging are becoming more prevalent. It is possible that in time conventional one-way broadcasting systems with no interactive capability will slowly become passé. This problem will be discussed in more detail in the text that follows.[28]

In this process of moving toward integrated and interactive systems, there are two extremely important trends. One trend is the movement to create hybrid telecommunications networks that combine fiber optic cable systems with wireless systems. There are at least three major examples of this hybridization. One is the combination of wireless mobile PCS systems with cable television systems or the largely wire-based Public Switched Telecommunications Network (PSTN). Here the wire, cable, or fiber systems act as node connectors. The second key hybrid system would be in combining or interconnecting wireless PABX or LAN systems in buildings with the traditional wired telephone network. This would typically involve using fiber optic cables as the vertical risers within the building structure but using wireless within the offices. The third case would be the use of wireless Local Area Networks (LANs) to connect to large and geographically extensive enterprise networks, or WANs, which are often "wired" or fiber networks. There will be more examples in the future as new technical developments such as even more advanced low earth orbit satellites and infrared bus systems become available and offer new service capabilities in terms of hybrid interconnections.[29]

The second trend, as noted above, is the move toward interactivity. This is a broad-based trend for all types of telecommunications systems, regardless of their purpose (i.e., data communications, voice, entertainment, imaging, etc.). The objective is to make telecommunications systems more versatile. This means enabling telecommunications systems to be more mobile, more quickly restorable, more flexibly controlled and utilized, as well as more intelligently managed. The recent decision of American Airlines to redesign their SABRE™ electronic passenger reservation system to allow for wireless access in order to adapt to overload or emergency situations is a case in point. Now SABRE networked laptop computers can be hooked up to designated airport gates to allow multiple access by reassigned or on-call service representatives. This has already proven effective in dealing more efficiently and urgently with short-term or crisis problems. This type of flexible and mobile response systems are indicative of future trends with wireless networks.

Many market analysts have concluded that traditional over-the-air entertainment broadcast entities are in danger because of their slow rates of innovation and particularly because of their low level of interactivity. Broadcasters have in many cases lagged in recognizing the need to move toward systems that are interactive, switched, and intelligently and flexibly controlled. In short, this lack of innovation suggests the traditional broadcasters may be in danger of becoming technologically obsolete. Other analysts, however, suggest that it only means that traditional television broadcasters may simply concentrate their efforts toward programming and phase out of distributional networks. Critics such as John Barlow have accused network executives as having a bomber pilot's mentality of a one-way dropping of "information bombs" on the consuming public.[30]

This same type of concern also applies to the newspaper industry. Some have envisioned newspapers in an interactive mode of operation. The Knight-Ridder newspaper chain and other newspapers are sponsoring research into the concept of twenty-first century electronic newspapers. In this view of the future newspapers would be electronically published on portable electronic tablets that could be updated by wireless radio networks. Such a lightweight and portable unit could be carried from room to room or even on a bus or train. Thus, the future electronic highway in this wireless-driven approach could be more flexible, accessible, and portable than your living room television set.

A third and final key trend is in the move to create some form of universal protocol and transmission scheme that is truly open to all forms of telecommunications regardless of the application. This trend is today being driven by digital transmission systems but without any clear winners. The options for data services today still start with an increasingly obsolete and slow X.25 system. They do, however, move progressively upward to Transmission Control Protocol/INTERNET Protocol (TCP/IP), frame relay, then Switched Multi-Megabit Digital Service (SMDS/DQDB), and ultimately to cell relay. Cell relay is just beginning to be operationally expressed in the format of Asynchronous Transfer Mode (ATM) networks both on private and public systems. This new high throughput, flexibly routed ATM system can be utilized to provide all forms of data, voice, and video services in terms of broad band ISBN networks.[31]

The problem is that today the systems represented by broad band ISDN and ATM require too much bandwidth to travel through most wireless systems. Only broad band satellites operating in the SHF band can handle gigabit per second traffic. The simpler and lower-speed protocols can and do work on wireless systems, but they are in a sense being phased out in favor of the higher-efficiency and higher-speed protocols that have often been optimized for fiber-based systems. In this sense, wireless systems must either increase their throughput performance or find better ways to interface effectively with the coming broad band ATM systems that are designed for optical systems. The field trials with millimeter wave-based LMDS wireless services that are now being carried out in the United States as well as experiments with office-based infrared LANs, PABXs, and multipurpose "buses" are the forerunners of new wireless systems that are broad band enough to be compatible with ATM systems. In fact, for intraoffice and short-range applications, there are already operational, off-the-shelf wireless systems that can operate at very high bit rates. These, however, can only provide very localized service and often have special requirements such as mirrors or lenses for concentrating the infrared beams.[32]

The key technical trends in the field of wireless telecommunications that will be addressed in this book include the following areas:

(a) New digital encoding and modulation techniques that can, among other factors, increase quality, reliability, tolerance to interference, throughput, and cost efficiency. (The issue of protocols and transmission standards, especially in the context of B-ISDN services will also be considered in this respect.)

(b) Digital compression, which can increase cost efficiency and "effective throughput" with power- and frequency-limited systems.

(c) Frequency expansion strategies that include new higher band allocations and more intensive frequency reuse techniques.

(d) Use of infrared frequencies for expanded wireless services.

(e) Use of new types of satellite constellations and orbits for new wireless and mobile applications.

(f) Advanced concepts involving hybrid wire and wireless systems.

The technology section of this book, presented in Chapter 4, introduces the basic concepts and applications that can flow from the use of the above technologies. These chapters will not only explain the various technologies but also seek to analyze how they can be expected to interrelate and produce systematic gains in the field of wireless telecommunications in coming years.

1.7 POLICY AND REGULATION IN WIRELESS TELECOMMUNICATIONS

Key to understanding policy and regulation in the wireless domain is the clear understanding of how "wireless systems" differ from the so-called "wire" systems. The historical separation of wired and wireless telecommunications regulation and standards-making goes back almost a hundred years to the invention of radio transmission by Guglielmo Marconi at the turn of the twentieth century and even further to the invention of the telegraph some sixty years before that time.

The International Telegraph Board was first established in Bern, Switzerland and then later moved to Geneva, Switzerland to become the International Telecommunication Union (ITU). This organization was established in 1865 in order to regulate the new electronic messaging technology invented by Samuel F. B. Morse in the 1840s. The International Telecommunication Union (ITU) was thus firmly established for some five decades based on wire technology before radio transmission first emerged in the first decade of the twentieth century. The result of this new technology within the ITU structure was to create for the purposes of standards-making two bodies, both known as International Consultative Committees. One unit of longstanding and proven worth was for telephone and telegraph and the other new and "exotic" or "experimental" unit was for radio technologies. In French, these acronyms thus became the "CCITT" and the "CCIR." This formal separation of the two technologies at the international level was mirrored at the national level as well. Only in the new structure of the ITU just implemented is this division between wire and wireless technologies merged together for standards-making purposes.

It is only recently that the concept of integrated digital services and seamless international telecommunications networks suggested that a separation might be unwise. Further, the idea that radio is somehow the new or not entirely proven telecommunications technology still remains within the fabric of the international telecommunications community. When the ISDN concepts were developed within the CCITT, the radio and satellite people were largely consigned to an observer status.

As only one example of this problem as of the late 1980s is the case of high-speed ISDN codecs standards. From 1986 to 1989 INTELSAT and COMSAT spent millions of dollars to develop a broad band satellite system designed to operate at 140 Mb/s. Then in 1989, they abruptly found out as their development was almost complete that the CCITT committee experts had decided that the broad band standard would be shifted to 155.5 Mb/s instead.

At the time the division was first created between wire and wireless almost a century ago it appeared to make a great deal of sense, but in today's world this division has been rightly questioned. The ITU as the entity responsible for international regulation and for standards-making in telecommunications has been reorganized. It will integrate the "wire" and "wireless" standards-making into a single unit, but not quite. It will still have a radio-telecommunications unit for frequency registration and coordination. It is still not clear how integrated the wire and wireless standards and policy-making process will really be. As the world of telecommunications migrates toward such basic concepts as Integrated Services Digital Network (ISDN), broad band ISDN, SONET, SDH, and ATM, the logic of integrated, interdisciplinary, and hybrid approaches to standardization and regulation seems to make a great deal of sense. Full applications of these principles is still incomplete even with the new reorganization of the ITU.

The problems and issues go well beyond the ITU. There are other international organizations who are working in the telecommunications standards area as well. These include the International Standards Organization (ISO), which is working with the ITU on the basic concepts of Open Systems Interconnection. In addition, there are the International Electro-

Technical Committee (IEC), the IEEE, and the General Agreement on Trade and Tariffs that will soon become the World Trade Organization.[33] When it comes to standards-making, the more units involved, the more difficult the process becomes.

Today, the development of telecommunications standards, despite the work of the ITU, the ISO, the IEC, the IEEE, and the GATT, has moved toward the national and regional level. This "regionalization and nationalization" of the process serves, in effect, to undercut the power and authority of the international standards-making agencies. In Europe there is the European Telecommunications Standards Institute (ETSI). In the United States there is the American National Standards Institute with specific units devoted explicitly to developing telecommunications and information technology standards. In Japan there is the Telecommunications Technology Committee, while other regions such as South and Central America have their CITEL Committee, Africa has the Pan American Telecommunication Union (PATU), and the Asia-Pacific Telecommunity considers standards issues as well. Often the pattern of standards evolution today sees several national or regional "proto-standards" evolve first. Then within the ITU framework and that of other international standards organizations the process of "standards coordination" replaces that of a single international standards development.

There are thus a number of formalized procedures that are directed toward developing systematic global standards. These activities include protocol conversions, open systems interconnection (i.e., OSI), transmission systems, switching and signaling systems, and a host of other areas as well. Despite these efforts there are now multiple international approaches to network design. Ironically there are now more and more standards with less and less effectiveness and yet they cost more and more to develop. There is a witticism that suggests that multiple standards are indeed hard to avoid in a competitive and highly political world. It goes as follows: "We love standards. That is why we have so many of them."

Unfortunately, today this observation is really not just an ironic thought but rather more of a reflection of reality. It is sadly and rather wryly true that even the concept of "openness" that is intended to make everything compatible with everything else based upon "universal standards" has at least three different names. There is the Open Systems Interconnection (OSI) of the ITU and the ISO, the Open Network Access of the FCC, and the Open Network Provision of the European standards group, ETSI. Despite these many standards-making efforts whose name begins with open, the idea of a truly open and truly universal standard has not made real headway. The TCP/IP standard used for the INTERNET system has in some ways made the most progress. The attempt to create a new version of TCP/IP for implementation may well have even undercut this standard in that the new version is not backward compatible to those using the older version.[34]

Despite the current moves to integrate radio and wireless telecommunications regulation and standards-making with the counterpart terrestrial "wire" activities, there are clearly important, objective, and performance-related differences between the two technologies that cannot be avoided or overlooked. These basic differences can be characterized in terms of: (a) spectrum use; (b) radiation or transmission of signals; (c) system architecture; and (d) reconfigurability, mobility, and other special service features.

Spectrum Use

The difference in the use of spectrum is obvious. Wire technology confines the transmission and its carrier waves within the medium that forms the conduit for communications. This allows the same frequency to be used as often as required, because simple insulation and "isolation" of the transmission frequencies within the "wire" can prevent interference even with immediately adjacent conduits. In the wireless environment, however, the frequencies are "used." This means they spill out into the environment and create interference for anyone attempting to use the same frequency. Clearly techniques already discussed such as physical isolation of satellite beams, cellular systems, polarization discrimination, coding, and other techniques allow frequencies to be reused many times or be used much more effectively between and among shared-use systems. There are, however, clearly limits. In short, there is a fundamental difference in standards-making, regulation, and resource allocation when it comes to wire and wireless telecommunications systems. Clearly wireless radio services just in terms of spectrum use alone pose a much more difficult problem.

Radiation. The issue of regulation of radiation of radio signals has several implications. One is simply to minimize interference into other telecommunications systems by simply limited irradiated power. There is the other concern of radiated power posing a health problem to the user of telecommunications equipment and networks. Both the technical and the health-related aspects of regulating electromagnetic radiation are considered of great importance today for operational, technical, financial and health-related reasons. Particularly in the microwave and millimeter wave areas, there is ongoing research that suggests that more stringent health standards may indeed be required. To date, however, ongoing empirical tests of radiate power and health risks are still inconclusive.

System Architecture. Many people would consider the system architectures and design features of wireless systems to be essentially technical or operational issues, rather than an area of policy or regulatory concern. In fact, this is one of many areas where technology and policy intersect. Telecommunications networks that are hierarchically concentrated with fixed or wire-based links and designed to funnel increasingly heavy streams of traffic to and through high-capacity switches have been considered the "normal" form of communications for many decades. This reflects not only telecommunications tradition but also a traffic engineering concept that evolved with so-called national telecommunications monopoly organizations. It has been thought for more than a century that the larger such concentrated hierarchical networks become and the more traffic they service, the more naturally efficient they will be.

Today new options are available. Wireless networks can clearly mimic the architecture of the traditional hierarchical or "vertically integrated" network, but they can also create "horizontal" networks that "bypass" the concentrated network through "mesh" or webbed networks that connect remote to remote nodes directly together. The basic differences in traditional "vertical" or concentrated networks versus "horizontal" mesh networks are shown in Figures 1.9, 1.10, and 1.11.

The key to note is that up until the 1980s most regulatory authorities and most telecommunications engineers thought in terms of large natural telecommunications monopolies

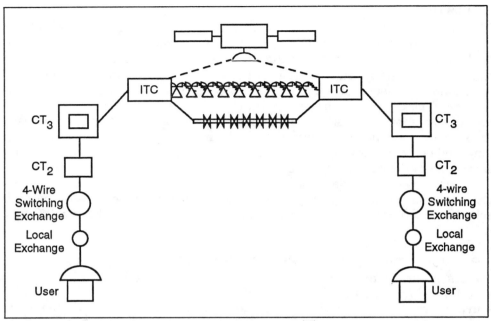

FIGURE 1.9   Hierarchial Telecommunications Network[35]

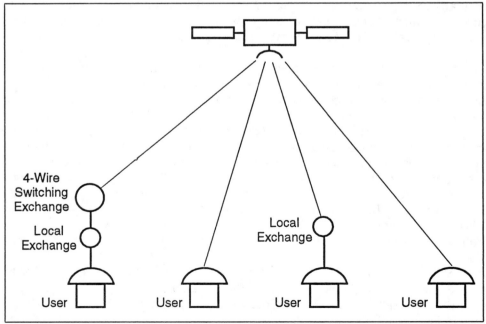

FIGURE 1.10   Satellite Horizontally Linked Network[36]

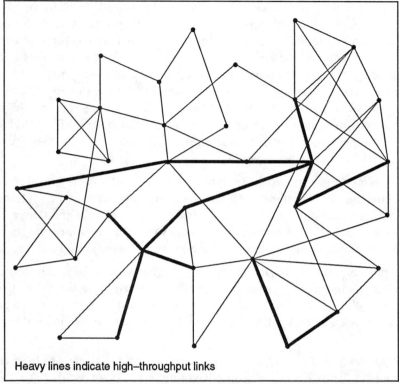

Heavy lines indicate high–throughput links

FIGURE 1.11   Routing Flexibility of Mesh Architecture in Telecommunications[37]

and economies of scales achieved through concentrated telecommunications networks. This view also saw telecommunications as being essentially rooted in wire and switches. It was the "other services" such as broadcasting entertainment, emergency and military services, taxi cab and truck fleet operators, and amateur radio "hams" that were seen as the users of wireless technology. By the mid-1980s this basic paradigm that constituted conventional telecommunications began to crumble. The neatly sorted out and well-compartmentalized world of telecommunications began to shift and then, in effect, to seriously unravel. Wireless, which received an impulse in terms of long-distance communications with the advent of satellites in the 1960s and 1970s, rather quickly came of age in the world of telecommunications with the cellular radio telephone in the 1980s.

Wireless communications grew and expanded everywhere in the 1980s. Beyond the most obvious case of the cellular telephone, there was also Instructional Fixed Television Service (IFTS), wireless cable, Satellite Master Antenna Television (SMATV), Direct Broadcast Satellite, mobile and fixed satellite, and specialized mobile radio; even private wireless service for LANs and PABXs began to be taken seriously. Ironically it was when fiber optic cable was seemingly exerting its dominance in high-capacity transmission systems that wireless actually made its major surge forward.[38]

This wireless revolution in telecommunications by almost perfectly overlapping with the divestiture of AT&T, the wide spread growth of fiber optics, and a worldwide surge toward privatization, deregulation, competition, and liberalization is sometimes overlooked. Today, however, policy makers are beginning to recognize that wireless is not only important, but it is, in fact, a key "enabling" technology that is allowing many of the regulatory innovations to succeed. The single most prominent industry move in the United States in this respect, namely AT&T's acquisition of McCaw Cellular, has certainly highlighted the issue. The moves into strategic partnerships by Air Touch and US WEST and by Bell Atlantic and NYNEX represent moves of comparable scope. There are, however, dozens of smaller yet still significant examples as well.[39]

Wireless services were an easy way to bring new capital and new competitors into the new arena. The ability of wireless to either overlay or expand existing systems or alternatively to compete with established systems without massive new capital investments is actually a key element in new competitive regulatory environments around the world. With wireless telecommunications the basic ground rules of the entire industry are changed.

It is now possible to have multiple carriers, alternative system architectures, and competitive environments without necessarily creating over-investment or placing one's future on a single technology or media. Most significantly, at least in the U.S. competitive context, it is possible to have true competition in the last mile of service at the local loop level.

Reconfigurability and Mobility.   The cost and the cost efficiency of fiber optic cable is indeed impressive. The idea that silicon-based fiber is as cheap as "dirt" or at least sand is certainly a compelling idea although somewhat misleading claim made in favor of advanced cable systems. Free-space transmission, however, is dependent upon no media except the earth's environment or in the case of satellites, the void of space. This is indeed a very cost-effective medium, although the move to auction off exclusive frequency allocations in the United States, New Zealand, and elsewhere does suggest this "free" access to the electromagnetic environment is also likely to change around the world.

The true beauty of wireless communications is not its ability to use a nearly free medium for transmission, but rather the instant ability to connect virtually any point to any other point whether stationary or moving, whether on the ground, sea, or sky, or whether in an office, home, or car. Fiber cannot do these things, nor can it be instantly reconfigured to add or subtract points of service. This fundamental difference between wireless and wired service gives rise to much different regulatory concerns.

Fixed wired networks are static and thus easy to monitor. Furthermore, wire networks typically do not irradiate power and frequencies into the earth's environment. Wireless systems in contrast are dynamic, always changing, always prone to create new or unexpected interference into other networks, and are possibly able to create health problems or concerns. The very strategic advantage of wireless communications, i.e., their mobility and reconfigurability, also makes their effective oversight and regulation much more difficult.

1.8 THE FUTURE OF WIRELESS TELECOMMUNICATIONS

Market demand, new services and applications, new technology, and policy and regulation will continue to create an exciting future for wireless telecommunications. Some services will peak and ebb, but other new services will fuel continued rapid growth.

Key aspects of that future will be defined by new standards and by new business practices. Other key factors will be new innovations that allow wireless services to be even more mobile, broader in band-width, and of higher quality. Part of this trend will be that wireless services will increasingly add an entertainment- and consumer-convenient component in addition to its current high level of focus on business services. Finally, financial and tariffing matters and new forms of private networks and value-added services will define the other dimensions of this media's growth.

Attempts to project the future in many instances fail because they are heavily technology based. This is logical in the sense that technology often follows a reasonably predictable developmental path that can be projected in a linear or smooth path forward. Policy and regulatory patterns as well as business and economic factors are much more random and unpredictable. Nonlinear events are certainly much more difficult to understand, describe, and certainly to predict. In the majority of cases, however, it is the nontechnical factors that make the biggest impact on patterns of future development. This is not to say that the future cannot be intelligently considered as long as reasonable expectations are maintained about what is "knowable" and what is not.

Key factors that can be helpful with regard to the future are cycles of history and regulatory practice. One can also consider the vectors and speed of global change that may occur across the international landscape. Furthermore, parametric modeling can help show the level of impact that various changes in one area might make across the entire field of telecommunications.

Certain aspects of the future seem clear. The most obvious aspect is the likely growth and development of new and expanded mobile services with wider band applications. These may well include such aspects as tele-entertainment and expanded tele-location and tele-mapping services that may well accompany the creation of the concrete-based "intelligent highways" as opposed to the intelligent information highways.[40] The further evolution of tele-commuting and tele-work will also likely give rise to wider band wireless and mobile services. The prospect of low-cost, flexible, and highly capable "electronic tablets" that fall somewhere between the portable PC notebooks and the "toys" of Fisher Price and Mattel may give rise to totally new service concepts. These might include such fresh ideas as the totally portable electronic newspaper, which, as noted earlier is being developed by the Knight-Ridder Corporation, and the "electronic tutor" project, as is being designed by the International Space University and the University of Colorado. The band-width requirements and the commercial value of only these two new products/services could be prodigious.[41]

The future developments that are the least clear are the final results associated with the changing patterns of ownership and control of wireless services and products. The current framework of turmoil and transitional effects, frequently called "convergence," impacts every aspect of the field. At a minimum it spreads across the industries that we today know as communications, cable television, content (e.g., entertainment software, etc.), consumer electronics, and computers—the five C's. Figure 1.12 indicates this overall process, which has been called the "big bang." In such a process it is hard to predict the winners, the losers, or even the in-between "survivors" with any accuracy. The one known is that major changes will indeed occur.

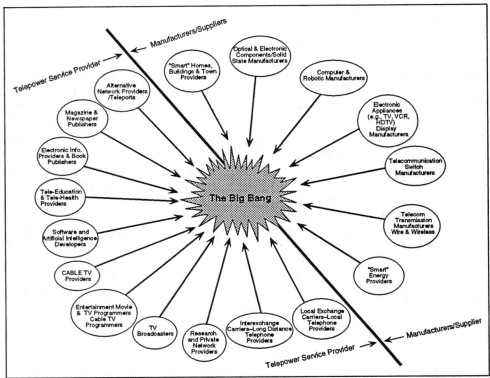

FIGURE 1.12   The Coming "Big Bang" in Information Technology Industries[42]

These industrial mergers, acquisitions, restructurings, new joint ventures and partnerships will likely have as much impact on the future as governmental reforms and regulations. This is because they will be so pervasive and fundamental. The application of wireless services to key social needs will also be a key factor in the future as well. Education, medical and health services, and public safety will define new tele-service markets, and many of these will require mobile and wireless services. This is not to say that new business applications and changes in governmental regulations and standards at the national and international level will not be important as well. The rich intermix of all these factors will likely create a rapidly growing wireless industry of the twenty-first century that will if anything place even more of a premium on the mobility and flexibility of communications services.

Despite all of the very positive aspects that indicate a very bright future for wireless telecommunications, it is always wise to recognize that in high-growth industries, change can occur rapidly and that it is advisable to be conservative in times of very high expectation and vice versa. There is also often a tendency to become over-enamored of new technology for its own sake as well as to overlook broad shifts at the societal level that can have unexpected impacts on particular industries when the dust clears. These may be baby booms, baby busts, shifts in commuting patterns, or new forms of credit-based spending.

The advice and cautions offered by Stephen Schnaars in his book *Megamistakes* with regard to predicting the future and anticipating market trends are well worth noting for any planner or market developer in the wireless telecommunications industry.[43]

1.9 EXPLORING THE COMPLEXITY OF WIRELESS TECHNOLOGY

The chapters that follow examine every major aspect of the field of wireless telecommunications in all its complexity—past, present, and future. They examine all aspects of the five drivers model—technology; applications and services; standards; economics; and finance, policy, and regulation, as well as the critically important marketplace. They attempt to present not only how and why wireless communications operates today but also to examine the critical dynamics of change, innovation, and institutional and political reform. Since the speed of change in this field, especially in the policy and standards areas is so rapid and since the complexity of interactions involved is also rather overwhelming, the key focus of this book is certainly primarily on the United States. Nevertheless, regional and international trends of importance, especially as they impact the U.S. wireless market, will be presented where and when relevant.

ENDNOTES

(1) Pelton, J.N., "The Globalization of Universal Telecommunications Services," in *Universal Telephone Service: Ready for the 21st Century* (Wye, MD: Institute for Information Studies); *Annual Review*, Aspen Institute, November 1991, pp.141-151.

(2) Meyers, R., ed., *The Encyclopedia of Telecommunications* (San Diego, CA: Academic Press, 1989); also see Newton, H., *Telecom Dictionary* (New York: Telecom Library, 1992).

(3) Deloitte and Touche, *1993 Wireless Communications Industry Survey* (Atlanta, GA: Deloitte and Touche, 1993); also see Ketchum, J., "Cellular Mobile Telecommunications" Tutorial on Wireless Technologies, Supercom 1991, Houston, TX.

(4) Ramteke, T., *Networks* (Englewood Cliffs, NJ: Prentice Hall, 1994) pp. 145ff.

(5) Keen, P.G.W., and J.M. Cummins, *Networks in Action: Business Choices and Telecommunications Decisions* (Belmont, CA: Wadsworth, 1994), pp. 48-50 and 82ff.

(6) Inglis, A.J., *Behind the Tube: A History of Broadcasting Technology* (Stoneham, MA: Butterworth Publishing, 1992).

(7) Deloitte and Touche, *1993 Wireless Communications Industry Survey* (Atlanta, GA: Deloitte and Touche, 1993).

(8) Ono, R., "Multi-dimensional Analysis of the Global Telecommunications Gap," unpublished Ph.D. dissertation, University of Hawaii, 1994.

(9) "NASA/NSF Panel Report on Satellite Communications Systems and Technology," (Baltimore, MD: International Technology Research Institute, July, 1993); also see "DBS Comes of Age: Direct-to-the-Home Entertainment Services Open Up A New Lane On the The Information Highway," *Uplink*, Summer 1994, pp. 4-7.

(10) Williams, F., *The New Telecommunications: Infrastructure For the Information Age* (New York: The Free Press, 1991).

(11) DeSilva, E.W., "PCS/Wireless Regulatory Update," *ICA Expo*, Dallas, TX, May 1994.

(12) Manuta, L., "Riding the Spectrum Wave," *Satellite Communications*, July 1994, pp. 24-26.

(13) Taylor, L., "PCS Frequency Auction," *Signals*, November 1993.

(14) Negroponte, N., "Digital Networks," *Telecommunications*, April 1993, Vol. 11, No. 4.

(15) Pelton, J.N., "Five Reasons Why Nicholas Negroponte is Wrong About the Future of Telecommunications," *Telecommunications*, April 1993, Vol. 11, No. 4.

(16) Deloitte and Touche, *1993 Wireless Communications Industry Survey* (Atlanta, GA: Deloitte and Touche, 1993), "Wireless Markets for the 1990s," Delphi Survey, 1994, ITP.

(17) *Global Market Forecasts for Telecommunications* (New York: Comquest, Nielsen Co., 1993).

(18) U.S. Congress, Office of Technology Assessment, *Critical Connections: Communications for the Future* (Washington, DC: OTA, 1990).

(19) Frieden, R., "Wireline vs. Wireless: Can Network Parity Be Reached?" *Satellite Communications*, (July 1994); also see Manuta, L., "Riding the Spectrum Wave," *Satellite Communications*, July 1994, pp. 24-26.

(20) "Future Applications for Extreme High Frequency Bands," *ACTS Quarterly*, Jet Propulsion Lab., September 1993.

(21) MacDonald, V.H., "The Cellular Concept," *The Bell System Technical Journal*, January 1979.

(22) Calhoun, G., *Digital Satellite Communications* (Norwood, MA: Artech House, 1987), p. 380.

(23) Cellular coverage for low earth orbit cellular reuse pattern—Iridium—Figure 1.8, p. 38.

(24) Pelton, J.N., *The How To Book of Satellite Communications* (Sonoma, CA: Design Publishers, 1992).

(25) Calhoun, G., *Digital Satellite Communications* (Norwood, MA: Artech House, 1987).

(26) "Advanced Digital Compression Produces Ten Fold Gain in Performance," *Uplink*, Fall 1993.

(27) Lang, R., and J. Sauer, "Scalable Dense Wave Division Multiplex Photonics for an All Optical Network," White Paper, Spectra Diode Laboratory, San Jose, CA, 1994).

(28) Ducey, R.V., and M.R. Fratrik, "Broadcasting Response to New Technologies," *Journal of Media Economics*, Fall 1989, p. 80ff; also see Markus, M.L., "Toward a Critical Mass Theory of Interactive Media: Universal Access, Interdependence and Diffusion," *Communications Research*, October 1987, pp. 191-210.

(29) Pelton, J.N., "Toward a New National Vision of the Information Highway," *Telecommunications*, September 1993, Vol. 11, No. 9.

(30) ———, "Toward a New National Vision of the Information Highway," *Telecommunications*, September 1993, Vol. 11, No. 9.

(31) Keen, P.G.W., and J.M. Cummins, *Networks in Action: Business Choices and Telecommunications Decisions* (Belmont, CA: Wadsworth, 1994), pp. 269-280 and 77-80.

(32) Shea, T., "Need for Broadband ATM Compatible Wireless Systems," Seminar Presentation, Interdisciplinary Telecommunications Program, University of Colorado at Boulder, April 1994.

(33) Codding, G., "The Reorganization and Restructuring of the ITU" (London: International Institute of Communications, 1993).

(34) Keen, P.G.W., and J.M. Cummins, *Networks in Action: Business Choices and Telecommunications Decisions* (Belmont, CA: Wadsworth, 1994), pp. 277-280.

(35) Chitre, K., "Horizontal versus Vertical Telecommunications Systems," *ITU Journal,* September 1980, pp. 150-154.

(36) ———, "Horizontal versus Vertical Telecommunications Systems," *ITU Journal,* September 1980, pp. 28-30.

(37) Stuart, J., "Mesh Network Advantages," Small Satellite Conference, Washington, DC, February 1992.

(38) Wittman, A., "Will Wireless Win the War?" *Network Computing,* June 1, 1994, Vol. 5, No. 6, pp. 58-71.

(39) "AT&T To Take Over McCaw Communications," *New York Times,* April 5, 1994.

(40) Berger, J., "How Do I Get There: Navigating In The Smart Car Age," *Wireless,* May/June 1994, Vol. 3, No. 3, pp. 28ff.

(41) Pelton, J.N., *Future View: Communications, Technology and Society* (Boulder, CO: Baylin Publications, 1992), pp. 65-88 and 131-146.

(42) ———, "Toward a New National Vision."

(43) Schnaars, S., *Megamistakes* (New York: MacMillan, 1987).

CHAPTER TWO

THE WORLD OF WIRELESS: FUNDAMENTAL CONCEPTS AND KEY SERVICES

2.1 INTRODUCTION

This chapter, the last of the introductory section, seeks to address the basic aspects of what wireless services actually are. It also addresses what they may become in the near and medium term future. Up-to-date information about how each service actually operates in terms of technologies and frequency allocations is also presented. Finally, it provides some background and history concerning these services. In the next section of the book the detailed aspects of wireless communications will be explored. This in-depth exploration will begin with markets and services, then turn to policy and regulatory matters and then finally investigate the key aspects of wireless technology.

2.2 PAGING AND POSITION LOCATION SERVICES

The paging service is one of the oldest and best established commercial wireless services. Its basic purpose is to let a subscriber in a distant or mobile location know that someone wishes to contact them. It is a low-cost version of having a portable telephone and is, of course, dependent on the person being paged being close to a telephone in order to quickly reply. If the individual being paged does not have access to a telephone, in say a remote location, then the value of the pager is greatly depreciated. Paging is evolving, however, to become an interactive messaging service as well.

Position location, or radio determination service, on the other hand, is designed to allow people at remote locations to know immediately exactly where they are through a process known as "map matching." This allows users to see their location on a display screen or to let others know where they are on a remote screen display. This position location service is important to fleet operators or other transportation or message-oriented firms with large mobile operations. This service, often called "fleet management," lets such users know where their trucks, buses, and delivery or service vans are currently located in virtually real time. Initially this position location or radio determination service was predominantly based upon "Doppler shift" calculations, but with the advent of the Global Positioning Satellite (GPS) system, operated by the U.S. Department of Defense, the GPS triangulation calculations now provide greatly enhanced precision. Today the single most important use of GPS is probably for aircraft navigation purposes. In fact, the U.S. Federal Aviation Administration (FAA) may soon certify GPS systems for Category IIIA and IIIB auto-landings for low-visibility conditions. Its use for ground and sea transportation is also increasing rapidly. Enhanced services that go beyond simple position determination are now beginning to become available. These are land-based GPS systems that are able to utilize so-called intelligent vehicle highway systems. These include the following: (i) best route calculation; (ii) travel advisory of new conditions potentially affecting the travel path; (iii) address matching; and (iv) inventorying, which allows specialized tracking, for instance, of a stolen or missing vehicle.[1]

In the case of paging services, there are actually a range of services available ranging from one-way page to interactive messaging. These start with a basic beeper or vibrator alarms that signal to the subscriber that they are wanted. More sophisticated versions of paging services include the ability to send short digital messages for readout, typically on a liquid crystal display. Even more sophisticated beepers actually send short digital voice messages or can transmit or receive faxes.

Although most beeper services are localized to a particular geographic area such as a city, there are also very sophisticated versions of paging services that operate from geosynchronous satellites over national or even international service areas. These satellite-based systems obviously not only provide much wider coverage, but also involve the use of available paging frequencies over a much broader area as well. The new national, regional and local narrow band PCS services will greatly augment capabilities in this area. Also, in the future, very sophisticated coding systems will be used for satellite paging service in the new narrow band PCS and conventional paging frequencies in order to achieve new addressing efficiencies. In short, high-efficiency coding systems with individual identifiers for thousands of subscribers can be used to compensate for the "broad scale preemption" of the narrow frequency bands assigned for this service. Some of these "smart" coded devices can even offer position location services as well.

The allocation of the new narrow band PCS frequencies with enough frequency to allow extensive interactive messaging is expected to help stimulate the market. The 3 MHz of frequencies allocated in three 1-MHz bands at 900, 930, and 940 MHz, will offer a great many options. This is because 11 national licenses are being made available for all 50 states of the United States. In addition there are ten licenses for regional areas known as Major Trading Areas. Finally, there are in fact another ten licenses in each of 492 local areas designed by the FCC as Basic Trading Areas. These licenses are typically for 50-kHz carriers, which are sufficient for interactive messaging.[2]

In the new environment of PCS narrow band services subscribers may be assigned a unique code. As the demand for frequency increases in parallel with an increasing subscriber base and especially the demands of interactive messaging, the need for advanced coding of services and cellular divisions will also likely increase. For national and international satellite-based paging systems like SkyTel™ using existing narrow band frequencies, there will be a challenge of matching the capabilities of the new highly capable narrow band PCS systems. The fact that the bidders for narrow band PCS put up almost a half billion dollars for this now highly precious spectrum suggests that these corporations believe that there will be a highly lucrative market in this area. In light of the high stakes it will also likely be a highly competitive one as well.

It should be recognized that paging and position location services do not necessarily need to be linked together. In fact, there are at least three distinct options. One can obtain paging service or enhanced paging services with interactive messaging involving fax, voice messaging, or electronic display. One can also obtain a readout or display of position location services at fixed or mobile locations. Or finally one can obtain both messaging and position location as a combined or hybrid service.[3]

Remote paging services with digital messaging can be optimized to send urgent messages immediately, but to hold routine messages, which will be sent when lower long-distance rates apply. Basic paging services currently start out at about $20 per month in comparison to cellular telephone monthly rates, which average between $70 and $100 per month. In contrast, position location normally involves free access to the GPS satellite system orbiting some 17,000 kilometers (or 12,000 miles) above the earth and thus only requires obtaining a mobile GPS receiver. A typical receiver for GPS navigation such as the Magellan™ hand-held unit is currently about $500 to $700 (US).

The process of position location works very differently from that of paging. In the case of paging, a one-way message is sent out to subscribers by a terrestrial transmitter or via a satellite relay. In the case of position location service, the process usually depends on reception of either a terrestrial or satellite signal. This can be accomplished by one of two techniques. It can detect a Loran C transmission, which allows calculation of an approximate location, typically to within a 800 to 2000 meters accuracy. Alternatively, with a Global Positioning System (GPS) receiver, one can receive multiple transmissions from the 28 GPS satellites now in medium earth orbit. Through triangulation calculations, the GPS receiver can determine the mobile vehicle's location to within a few meters tolerance. Other satellites operated by other countries, particularly Russia's Cosmos satellites, are also available for this purpose, but currently most commercial systems rely on the GPS system.[4]

There is a politically and militarily sensitive aspect to the position location service, and that is the very precise capability of the GPS triangulation system. This system was, in fact, developed for the initial purpose of precisely targeting nuclear missiles and today's "smart" cruise missiles. There were sufficient concerns over the potential "targeting" use or rather misuse of GPS technology, that the U.S. Department of Defense attempted to create a "fuzzy" algorithm for civilian positioning systems while retaining the more precise military application. This strategy was not entirely successful in that alternative ways using multiple calculations were devised to obtain an accuracy to within centimeters. Today a hand-held GPS receiver unit with map locator functions can be obtained that can indeed provide pinpoint accuracy. Shortly there will even be wristwatch-sized devices that can do the same. Information from three GPS satellites must be received to determine a two-dimensional fix of longitude and latitude, and to determine altitude one must have information from at least one additional satellite.[5]

The official name for this positioning or location service under the International Telecommunication Union's definition is the Radio Determination Satellite Service (RDSS). What is common to the paging service and the position location service is that only a receiver is required. Neither the paging device nor the GPS receiver transmit a return signal. Several manufacturers, however, are considering the development of integrated transceiver devices that can receive both pager and GPS signals, and also provide interactive voice and data services. At this stage this would require operation at multiple frequencies, but in the future integrated services may be allowed on multipurpose frequency allocations. This is but one example of how frequency allocations and coordination procedures developed out of historical practices wherein various uses were clearly and discretely differentiated, but that modern practices are pushing toward an integrated service.[6]

2.3 DISPATCHING SERVICES, FLEET COMMUNICATIONS, IMTS, AND SPECIALIZED MOBILE RADIO

For years the prime requirement of mobile communications, other than military, police, and emergency communications services, was to assist in various commercial operations such as taxi cab companies, trucking fleet operations, and messaging and delivery services. The various radio systems developed for this purpose were designed to be low cost, serviceable, and yet still provide "acceptable" quality. Short messages using this dispatched mobile service operated on a "push-to-talk" transmit and receive capability were utilized because of the very limited band-width available for this service. It was essentially an everyone-use-the-same-limited-frequency free-for-all. As demand expanded, the quality of service deteriorated and demands for more frequencies were largely unanswered. The next step was an innovation called Improved Mobile Telephone Service (IMTS). This allowed mobile telephone users to dial their call party without intervention of a dispatch operator and for the first time gave an exclusive assignment of the call channel frequency. Thus for the first time there was at least some level of privacy in the conversation.

In time, new frequencies were added to this IMTS type service, better and higher-powered technologies were utilized and even satellite services (such as the now defunct Geostar) were added in the 1980s to provide short messaging services to trucking fleets. The principal point here is that the ever-increasing congestion of the fleet-type of communications and dispatching service created a serious problem. Furthermore, the new but parallel demand for mobile services on the part of a much more general public set the stage for a major change. The waiting lists for mobile telephone service in this environment were quite long. The result was a careful study of how to intensively reuse the frequencies for mobile services. This study carried out by the National Science Foundation and the FCC led to the creation of a new mobile service that used cellular radio and sophisticated frequency reuse techniques. This effectively expanded the available frequency for mobile communications by an order of magnitude. As new advanced digital techniques are applied in this area another order of magnitude increase in capacity can be anticipated.[7]

Today, in addition to the rapid spread of cellular radio telephone, the specialized mobile radio services are also in a state of transition from antiquated, low-technology, and inefficient fleet communications systems to much more modern systems that can increase the effective utilization of available frequencies by a factor of ten to even 20 times through a combination of frequency reuse techniques and digital systems. This trend toward both high-efficiency cellular and high-efficiency SMR services, known as enhanced specialized mobile radio (ESMR), is particularly prominent in the United States. Here several regional and nationally based specialized mobile radio (SMR) companies have acquired out-moded fleet management frequencies and updated the old systems with digital technology to create a highly efficient and high-quality operation. These ESMR operations include, COMM ONE™ and NEXTEL™ (formerly FleetCall). In fact, it has been recently announced that NEXTEL intends to acquire COMM ONE and integrate it and all of its many other acquisitions into a single national EMSR service.

Another way of describing developments over the last few years is that the distinction between dispatching and fleet communications and other mobile services has become blurred in recent years. Frequencies are simply too limited and valuable not to introduce intensive reuse through cellular antenna systems and modern digital modulation techniques.[8]

2.4 CELLULAR COMMUNICATIONS SERVICES

The increasing demand for new mobile services, the increasing congestion of existing frequencies, and the corresponding deterioration in service quality all thus led to a concerted effort to implement new technical solutions.[9] In both Europe and the United States there was focused research into the concept of creating lots of smaller cells rather than allowing single, broad coverage service areas all using the same frequency band. This concept of dividing a broad service area into much smaller units or cells had many clear advantages.[10] First, frequencies could be reused many times by creating a pattern whereby frequencies used in one cell were separated from one another by cells operating in other frequency bands. This pattern of cells using common frequencies within a cellular system are described in Figure 4.1.[11]

Second, the power levels used in cellular systems did not have to be as high to cover the smaller cell size as opposed to the much larger former service sectors, which typically covered entire metropolitan regions. Third, the problems of reflections and physical interference within the cell area, known as multipath, were much easier to overcome. This is because of the greater proportionate or relative height of the transmit and receive tower located in the middle of the cell. The larger the service area, the higher the tower must be due to simple laws of geometry.

The basic workings of a mobile telephone system in terms of the telecommunications service is largely defined by the cell base station at the center of the cell and the Mobile Telephone Switching Office (MTSO), which aggregates the mobile traffic from several cells and switches it into the Public Switched Network. It is at the base station where the telecommunications operations deviate significantly from the normal Public Switched Telecommunications Network (PSTN) routing and connection. The incoming mobile signals are captured at the base station and then forwarded to the MTSO, which services some three to five cells. Each cell base station and especially the cell controller sends the call out to the mobile subscriber on one hand. It also acts as a receiver from the MTSO as well as forwards the mobile call to the nearest MTSO. The MTSO performs the switching operations for all the cells it serves and is the connecting link to the broader world of PSTN communications. The MTSO is thus the primary interface between the mobile network and the conventional telephone system, but the base station serves as the secondary interface in translating the call into the RF signals for mobile transmissions. The MTSO typically hands off a mobile call directly into the Public Switched Telephone Network (PSTN). After the MTSO hands the signal off to the regular telephone network the cellular service functions just like a regular call except for special billing requirements. Today the conventional telephone network is typically the only viable way to interconnect the cells, but in the future this can and will change. In future cellular systems and especially in PCS systems it may well be that cable television systems, alternative network providers, or even electrical energy providers will be able to perform the critical node interconnection service.

When a call is made to a mobile subscriber, all the base station cell controller initially sends out is a coded signal notifying the subscriber that there is a call seeking them. Assuming the subscriber is within the service area, the mobile telephone will automatically call the closest MTSO to establish the connection. In short, the subscriber always "initiates" the call regardless of whether the subscriber is calling from a mobile or remote location or is being called. Thus the subscriber always has to pay the much higher mobile telephone service fee anytime a connection is made even if it is a wrong number or a "junk" call from a solicitor trying to sell something.[12]

The communication from the mobile telephone is sent by a radio transmitter that has a range that can be three to even eight miles depending upon the density of traffic within the cell. The mobile telephone operates off the vehicle battery or generator, or if away from this power source, from a small battery within the hand-held unit. These batteries hold sufficient charge for about two hours of service, but some thin silhouette, lightweight units have a capacity of only 30 minutes. The new lithium ion batteries now becoming available will serve to either make cellular telephones lighter in weight or extend their usable power without frequent recharges.[13]

The mobile telephone is technically a "full-duplex transceiver," which means a radio device that is capable of transmission on one frequency and receiving on another. Half-duplex radio transceivers still used in dispatcher and other fleet services can require that the same frequency be used for transmission and reception and use a "push-to-talk" system. This takes some practice to avoid sending and receiving at the same time. The most important frequency band allocated to analog mobile telephone services is the VHF band from 890 to 902 MHz, which is also shared with radio-location services.

Although mobile telephone service is very convenient, the quality is often impaired by a variety of factors. These include problems with: (a) effective hand off of the signal between cells; (b) multipath, shadowing, and/or physical interference; (c) narrow band channels; and (d) congestion of channels in traffic saturated cells. All of these problems are straightforward to understand and the introduction of digital technology and the addition of more cells can likely serve to improve future performance in all these areas. The additional frequencies that will be made available with high-tier broad band PCS will also help relieve congestion.[14]

The problem of hand off of the cellular signal as a vehicle moves from one cell to another is a matter of sensing the loss of signal and transferring the call to an approaching higher signal strength cell. The transfer of the call typically sounds like the clicking you hear when picking up a new extension within a house. There can be problems that complicate a smooth hand off. These can include traveling at very high speeds, being stalled in traffic right at a hand-off point or various forms of interference or multipath signal distortion right at the hand-off point. There is another type of problem that occurs when the subscriber literally drives out of a cell at the edge of service where no more cells are available. This can lead to fade or even abrupt loss of signal.[15]

In the future there will be several new options that will minimize the loss of signal. Along most of the interstate highways continuous cells will be created to maintain terrestrial mobile services.[16] In more rural and remote areas several satellite systems will be available to provide mobile telecommunications services. In fact, "parallel types" of satellite systems may also offer a national CD-ROM quality radio service as well. The systems that will provide radio broadcast services will include the American Mobile Satellite Corporation and Telesat Mobile Inc. (both being geosynchronous satellite systems covering North America) and a series of low and medium earth orbit satellite systems that may be able to provide radio broadcast services as well. These LEO and MEO systems will operate on a global rather than just a national basis and will likely include Motorola's Iridium™ Satellite System, Qualcomm/Loral's Globalstar™ Satellite System, TRW's Odyssey Satellite system, the Aries Constellation system, and INMARSAT's Project 21 or INMARSAT P system.[17]

There is an important service feature in mobile communications known as "roaming." This is the ability to switch your mobile radio system from one system to another and to have your calls rerouted wherever you might travel. This is a feature that can allow you to change service from your local cellular system, even to allow interconnection along interstate highways, to plug into other cellular systems, or to link into satellite systems if and when they are available. The key is in advising the mobile routing system that you are not at your normal cellular address but in fact are roaming to other systems. Ultimately, there will likely be a universal telephone number. This number will be uniquely your own. A worldwide intelligent signaling system will automatically forward your calls to you wherever you might be. Today, the routing system is not "intelligent," and thus you need to indicate your change of venue to the cellular call routing system to have calls properly forwarded.[18]

Cellular radio telephone service is not only convenient for the consumer, but it is also flexible and increasingly multipurpose. Today cellular radio service is much more than a voice service. Once the basic cellular telephone service is installed there are lots of additional options. One can install a fax machine or a modem for data and operate much as one would on an ordinary PSTN line. The fax- or computer-based modem can be installed by "hard wire" or even by an in-car radio link via a wireless modem that connects to the mobile line. This allows a user in a remote vehicle to connect directly to computer terminals around the country. Thus one can connect either to send or receive faxes, e-mail, or to obtain other data services. With new Cellular Digital Packet Data service, reliability and performance in terms of system availability and throughput can be further increased.

With a wireless modem one can either be in the car or even have access via a portable terminal within a reasonable proximity to the car's "wireless bus." In short, data, voice, or fax are all available via mobile cellular services today. Although voice is the overwhelming source of cellular revenues today, most cellular operators project that 25 to 30 percent of revenues will be derived from data and fax services by 1997. Since newer cellular systems using CDPD techniques are being developed to interleave data services via packet radio techniques, this is likely to be a very profitable service in coming years.[19]

There are two ways to operate cellular data services. The first way is to have a wireless modem connect to a cellular telephone and simply transmit the data over the open voice link. The charge is the same as if it were for voice service. The difficulty with this approach is that it is difficult to maintain the link without losing the signal due to cell switching or other interference on the narrow 3-kHz channel. The other option, which is just becoming available, is the digital cellular packet data service. This service sets up a special data channel out of spare capacity that becomes available when setting up and breaking down regular cellular service. It is optimized for data services, especially short, bursty messages. This service, which can be more expensive than regular data service over a cellular phone, should provide a higher grade of service with less downtime and repeat transmissions. Finally, when digital PCS service becomes available in the next few years it will provide an even better level of service and potentially even more attractive rates.[20]

Electronic tablet notebooks with light pen I/O devices, such as the Apple Newton 2 products, are among the latest options available for virtual office operations. Except for differences related to the cost of the additional in-car terminal equipment, these in-car mobile data services are all billed at the same basic rate of $.33 to $.50 per minute plus some recurring basic monthly cellular service fee. One can also install an antenna to boost performance or rely on the pull-out telescope antenna that comes with the mobile telephone unit. While all of these combined voice and data mobile services do cost money, the monthly fees of perhaps $75 to $100 per month are not excessive for business-related expenses.

For the cost-conscious cellular subscriber, the mobile telephone can, in fact, be used very sparingly—almost like an enhanced pager. This means a self-imposed discipline of using the cellular telephone for calling only in emergency or urgent situations. It also means telling incoming callers that they will be called back at the next available pay telephone. At the opposite extreme is the intensive user such as an attorney who uses the mobile telephone as a means of extending billable hours each day. Such an attorney may well bill at say $200 to $400 an hour for professional services and also may add the mobile telephone charges to the client's bill. In such cases, the mobile telephone is truly golden.

Realtors, sales representatives, professional sports people, politicians, reporters, and public relations people all literally use their telephones as mobile offices that are sometimes even called "virtual offices." The leveraged effectiveness of such professions is enormous and the cost of the service is often a very small percentage of the net value gain. The creation of a complete electronic office including voice, data, fax, and even imaging services is now a real possibility for professionals who use their car as a true remote office. Today, a complete package that would be considered the equivalent of a virtual office contains many components. It would likely include a cellular radio transceiver, a reasonably high-speed modem, a fax printer, a notebook computer and display screen, plus software to integrate the "virtual office" together. This complete package can be purchased for about $6000, and the entire system can fit into a single briefcase.

In later discussions, the issue of how large this "high-value-added" subscriber base might become will be considered and evaluated. This will describe and analyze what a full set of virtual office capabilities might be and when such intensive-use market patterns might eventually level off. As of early 1995, however, these services were still growing rapidly. About 40 percent of the users generate about 90 percent of the revenues for cellular services. The same general pattern can be anticipated for the PCS market as well.

In the United States, the world's largest single mobile telephone market, this cellular service has grown in a decade from nil to about 16 million users and is still growing fast. Of these users about 1 million were using the digitally enhanced IS-55 standards that employ TDMA. Worldwide there are some 28 million cellular users, of which 4 million are digital (i.e., IS 54 largely in the United States and GSM largely in Europe); see Figure 2.1.[21]

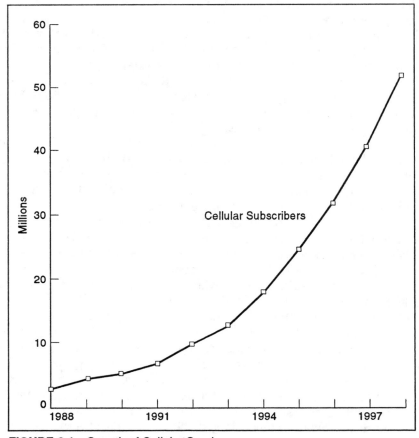

FIGURE 2.1   Growth of Cellular Services

The expected continued growth rate in North America, Europe, the Asia-Pacific, and elsewhere is certainly expected to be sufficiently strong to fuel the next tier of mobile telephone growth. This second tier of wireless telecommunications growth, which will come in the latter part of the 1990s, is represented by Personal Communications Service (PCS). Unlike the AMPS standard, which is based upon Frequency Division Multiple Access-based (FDMA) telephone service, this new PCS service will be either Time Division Multiple Access-based (TDMA) or Code Division Multiple Access-based (CDMA) digital cellular. Standards-making efforts in Europe have centered on the TDMA based GSM Standard. But in the United States the results are much more complicated. The Telecommunications Industries Association, the Electronic Industries Association, and ANSI through the so-called Joint Technical Committee ended in late 1994 with a compromise of adopting multiple standards that allow the use of both TDMA and CDMA systems. This United States standards making effort includes some five high-tier standards that include TDMA and CDMA standards plus two low-tier standards as well. The details of this are provided in the Standards chapter later in this book.

This new second tier or phase of mobile telecommunications market growth is projected to be strong enough to support the investment of billions of dollars in these new services. Furthermore, it is important to understand the degree to which analog services will also continue to be an important part of the market well into the mid- to late-1990s. Opinions, however, are mixed in this respect. Some clearly support the early maturity, enhanced performance, and improved cost effectiveness of digital TDMA and CDMA technology versus the more conventional and "lower tech" analog FMDA technology. Others support the better-proven analog technology. In short there will be continuing tension between analog and digital systems. Even so, many major operators, such as Air Touch, McCaw, and Bell South, will enter both technologies and, in effect, ride both horses.

Ameritech, with over 700,000 subscribers and therefore one of the larger cellular providers, recently announced that extensive field trials with digital cellular service (i.e., TDMA) found that performance was more unreliable than analog services and that they were therefore deferring implementation of this digital technology for the time being. While one can be skeptical of these results because of their very large installed base in analog equipment and the fact that there are about 1 million seemingly satisfied TDMA users under the existing EIA and ITA IS-55 standard, one must not be too quick to write off well-proven installed mobile radio systems. The issue of the relative merits of TDMA versus CDMA operations is addressed in more detail in Chapter 4.[22]

2.5   DIGITAL CELLULAR AND PERSONAL COMMUNICATIONS SERVICE (PCS)

The logic that suggested that cellular mobile telephone service might be a good idea is now being invoked in favor of what is heralded as an even better idea. This is, of course, personal communications service and digital cellular service. The key points of interest include defining exactly what is meant by these new services and understanding which frequency bands they will use. It is also important to investigate the various new standards that are being developed and implemented and to determine how they relate to conventional analog cellular telephone services in terms of performance, cost, and service capabilities. In summary terms, digital cellular communications provides significantly expanded telecommunications capacity for the same amount of frequency space.

PCS further provides higher quality of service and requires less power to operate and/ or lower-gain mobile transceivers. The shift to digital technology also represents several new opportunities. One significant change is the allocation of new frequency bands for this service, even though the IS-54 digital service uses the existing 800 MHz band. Another opportunity is to operate with much smaller cell sizes, called microcells or even picocells. The speed of implementation of microcellular technology is far from clear since there are many trade-offs involved. This new personal communications service is being allocated in three different ways that include national, regional, and local urban coverage areas. In addition there are unlicensed network equipment systems and wide band voice and narrow band messaging allocations. The wide band systems will be allocated in 10 to 30 MHz bands while the narrow band systems will typically be 12.5 to 50 kHz. As a totally new set of services there are many standards and frequency allocation issues still to be resolved.[23]

On the plus side of PCS service is the ability to operate at much lower power levels for the cell base station and the individual mobile transceiver. Furthermore, much more intensive frequency reuse can be achieved with up to ten times more channel capacity being derived from the same frequency allocation.

On the negative side, there may well be ten times more nodes that must be interconnected to provide service. Accordingly, the switching architecture of the system becomes much more complex. Also, the very small size of the cells and the very low power levels mean that digital PCS or microcellular service can be provided only to pedestrians or fixed locations, but not necessarily to vehicular traffic. This is because higher power is needed to reach high-speed vehicles especially when moving through trees and high rise urban environments. Recently, designers of PCS systems have been evaluating higher power and larger cell designs in order to provide mobile services to automobiles and other vehicular traffic.[24]

Even more of a problem is that of rapid and almost constant hand off between the microcells that a car might pass through every few seconds. Since today cellular service is predominantly provided for vehicular mobile traffic, this is a major deterrent to the use of microcells for this purpose. Instead, the new digital PCS services will perhaps tend to resemble conventional cellular services but will be offered with the full advantage of new frequency bands and advanced digital modulation and compression techniques. Figure 2.2 illustrates the basic concept which is that as the number of cells increases and as the cell size decreases, the number of node interconnenctions required for Personal Communications Services will go up exponentially. This major consideration in designing the PCS architecture is also a key factor in planning by cable television organizations who foresee the opportunity to provide node interconnection services. The available options for digital node interconnection include, in addition to cable television corporations, local exchange carriers, cellular and ESMR operators, teleports, and alternative network providers who may have extensive local loop capabilities. The fact that shrinking cell size associated with PCS systems requires not only more transmitters, but an increasingly complex switching and trunking network is a major economic, technical, and operational constraint in the design and implementation of a successful, reliable, and effectively expandable system.

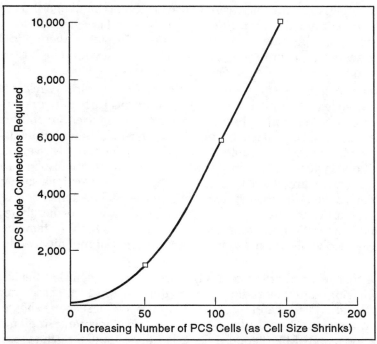

FIGURE 2.2   The Exponential Growth of PCS Node Interconnections

Another key consideration in PCS implementation is that related to the security of the mobile conversation or data transmission. Key steps that are being implemented to protect confidentiality are those set forth in Figure 2.3. It is anticipated that similar and perhaps even more stringent security measures will apply to the future mobile services such as IMT 2000 and FPLMTS.

Some PCS Security Issues
- All PCS Systems will be required to provide some level of voice and/or data encryption.
- IMSI, SID, NID, ESN, and any other ID or Authentication or billing information will be 'highly' encrypted.
- Many systems will employ a TMSI (Temporary Mobile Subscriber Identification) to provide user anonymity.
- This equipment is planned for export around the globe. Further, there is great pressure from the vendor community to roll back ITAR.

FIGURE 2.3   PCS Security Considerations[25]

2.6 WIRELESS PBX AND LAN SERVICES

A rapid expansion of wireless technology is expected inside modern business offices in the twenty-first century. This is because of the flexibility and cost savings the service can provide. As such wireless systems increasingly become broad band and thus capable of handling new multimedia and desk-top video, voice, data, and imaging service, the popularity of wireless services can be expected to grow. In many instances fiber optic cabling will be used to provide the basic riser telecommunications system within high rises, but the floor by floor or room by room service may well be wireless.

The reason for installing such systems is quite straightforward. Wireless telephone/private branch exchange (PBX) systems or wireless based local area networks (LANs) for computer communications networks provide total coverage of the entire building and even its grounds. Shifting an office, a desk, or a computer terminal can be done instantaneously with no rewiring, recabling, or change of number. For telephone service, total coverage by portable telephone is possible not only at desks or in offices, but also in hallways, closets, rest rooms, or virtually anywhere else. Although metal obstructions such as file cabinets can be a problem, a well-designed system can provide total flexibility and freedom at only modest cost premium and can provide a full range of service. Further there is also the option of overlaying an unlicensed wireless PABX or wireless LAN system over the top of an existing wire system to allow for expansion and the special mobility needs of certain key personnel.

If there is a concern of importance it might well be with regard to the long-term effects of the microwave or millimeter radio waves on the user's health. These systems have very low-power irradiation levels and extended health-related tests have shown such systems to be indeed quite safe. These systems have been developed and implemented in many large corporations, and IEEE Standard 802.3 is fully adopted. Such systems are thus no longer considered experimental and in fact are being widely installed as operational networks.

Figure 2.4 shows the type of wireless PBX/LAN system that can be installed today in a business office environment.[26]

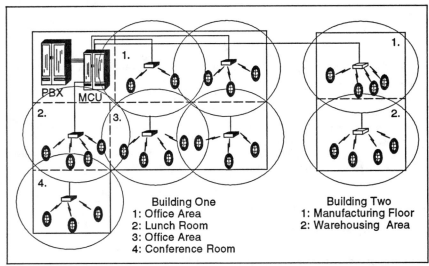

FIGURE 2.4   Wireless LAN and Wireless Public Branch Exchange Systems

There is, however, research and development underway to develop broader band and even more benign wireless systems for business offices that use the same infrared frequencies now used for remote controls of television sets. These systems will be briefly explained in the next section.

2.7 INFRARED BUSES

The wireless services described in the previous sections are currently in service. Wireless services in the infrared frequency band are today only in limited use for such purposes as remote controls of television sets and other appliances, garage door openers, and "remotes" for toys and specialty gadgets. There is the potential that these frequencies could be used for a wide range of business communications purposes. The very broad band capabilities of IR systems coupled with what appear to be very low health hazards from these systems actually make this application highly desirable.

There are, however, a series of difficulties to overcome. First of all, light, any spurious infrared source, and physical blockage can disable or disrupt such an infrared communications system. Second, there are no reliable and cost-effective infrared systems that can provide complex switching, bridging, or routing functions for office communications. Although these may be developed, such systems today are several years away from being fully realized. There are, however, some specific wireless LAN and Token Ring infrared-based short-distance transmission systems now available. Typically, these infrared wireless LANs need focusing mirrors to concentrate the IR signal between computers, routers, or bridgers. This is in order to ensure that the signals are sufficiently robust so that the signal is not interrupted or re-diverted by spurious signals. Thirdly, it should be noted that radio systems such as the Motorola Altair™ system are today highly reliable, extremely mobile, flexible, low cost, safe, and capable of providing sufficient throughput for most known applications.

Thus, the market for broad band infrared buses and infrared switching, routing, and bridging, which are highly reliable and designed for broad band services, will undoubtedly develop, but several years into the future. Since LAN systems are more precisely defined and route specific, as opposed to PABX operations, it is likely that they will be in the market somewhat sooner.[27]

2.8 SATELLITE SERVICES

It has been customary to refer to satellite telecommunications as a single generic service. The reason for this is largely historical. This is because satellites began offering one type of telecommunications service known as Fixed Satellite Services (FSS) as defined by the International Telecommunications Union (ITU). This service was typically offered by satellites in geosynchronous orbit and usually it was for long-distance or overseas services. Today, such a simplistic approach to wireless telecommunications satellite services is really not possible because of the rapid diversification of satellite types, satellite orbits, and satellite-based services.

The basic concepts and services for satellite systems will thus be presented as follows: (a) fixed satellite services from geosynchronous orbit; (b) elliptical and low to medium earth orbit satellites for voice communications; (c) elliptical and low to medium earth orbit for store and forward data services; (d) direct broadcast satellite services; and (e) military and national defense-related satellite services; and (f) other satellite services including intersatellite links, position determination, satellite geodesy, search and rescue, and data relay for scientific purposes and other satellite applications. There are, in fact, more than a dozen official space communications services as recognized and defined by the International Telecommunication Union (ITU). These are described in Chapter 5.

Some critics are, in fact, critical of this complex series of definitions. This is because the range of services provided by satellite keep changing, the markets keep shifting, and the dividing line between space and terrestrial communications services are increasingly difficult to discern. This has led to various initiatives to move toward a more flexible and less rigidly defined basis for frequency allocations to these services. Some have sought "multipurpose service" allocations that would allow several possible uses of the same band. Others have, in effect, found loopholes in the frequency allocations to provide expanded services by stretching the definitions.

Today allocations and sharing formulas for satellites should likely focus much more on the orbital characteristics than the specific service category. Unless service definitions are broadened and sharing guidelines extended on the basis of new digital coding technology, the problem of orbital frequency congestion will only tend to become worse.

The ASTRA satellite system has used Fixed Satellite Service frequencies to provide a service that seems very much like direct broadcast satellite services or, in ITU parlance, BSS services. In the case of ASTRA no technical or operation limits were exceeded in terms of power levels or other service constraints. INMARSAT provides mobile satellite services to oil drilling platforms even though the mobile terminals in this case are only moving at the rate of the continental drift. In general, lenient interpretations have probably served to stimulate industry growth and expansion.

2.8.1 Fixed Satellite Services from Geosynchronous Orbit

Today there are over 200 geosynchronous satellites providing fixed satellite services. The largest portion of these satellite systems are covered by the national systems of the United States and Russia and the various regional and international systems represented by the INTELSAT and INMARSAT global satellite systems, plus EUTELSAT, ARABSAT, and the private international systems like Orion and PanAmSat. Altogether there are nearly 30 countries with domestic satellite systems and many dozens of countries that lease spare satellite capacity from INTELSAT or domestic satellite systems for their long-distance telephone and television distribution services.[28]

There are three key bands that have been allocated to fixed satellite services for telecommunications. These are the C Band (6 GHz/4 GHz), the Ku Band (14 GHz/12 GHz), and the Ka Band (30 GHz/20 GHz). The specific frequency bands are provided in a comprehensive chart later in this chapter. The first of the paired frequency in each band is used for the up-link to the satellite, and the second frequency is used for the down-link. The higher the frequency, the greater the unwanted precipitation attenuation that is experienced. This serves to degrade the communications performance. Therefore, the lower band in each frequency pairing is always devoted to the down-link and the higher frequency to the up-link. Most satellite systems for fixed satellite services operate in geosynchronous orbit in order to achieve continuous 24 hours per day coverage of the desired service area and in order to work with ground antennas that can be continuously pointed to the same location in the sky.

Today's sophisticated communications satellites can achieve very high throughput capabilities of about 3 gigabits per second or the equivalent of about 120,000 voice circuits. This performance is obtained by using very-high-power spot beams to boost e.i.r.p. (effective isotropic radiated power). Other key capabilities that help to increase capacity include a variety of frequency reuse techniques, polarization discrimination techniques, and sophisticated modulation, encoding, and digital compression systems. In the future on-board processing, signal regeneration, and advanced space switching systems will further boost satellite capacity while also allowing operation to smaller, cheaper, and more conveniently sized ground antennas.[29]

Fixed satellite services are very well suited to provide telecommunications to rural and remote areas where there are low concentrations of traffic and to provide radio, television, or other one-way service from a single point to multipoint locations. Further fixed-service satellites are also well suited when a high degree of connectivity is desired among a large number of nodes. Although fixed satellite services can indeed be used for high-density telecommunications trunks, recent advances in fiber optic cable that increase digital throughput and quality while also decreasing costs have served to make cable links the preferred transmission route over longer and longer distances. Fundamentally satellites and fiber optic cable are complementary technologies as much as competitors.[30]

The traditional concept of how telecommunications systems should be designed involved the idea of hierarchical networks that are concentrated together through switches and then sent over long distances and then deconcentrated. This architecture as well as the alternative architecture now possible through advanced "intelligent" satellite system design was shown earlier in Figures 1.9 and 1.10.

There is now actually a quite different approach to designing telecommunications systems and particularly advanced satellite systems that is based on using a "mesh network" architecture. Here end-users or nearly end-users are simply connected together through a highly efficient space-based networking system. This type of on-board switching system is often known as a "bypass" system because it "leaps over" terrestrial switches and uses on-board processing in orbit to regenerate and send the signal directly where it is supposed to go. When satellites operate in this optimum mode rather than mimicking the ground-based hierarchical systems, they can be much more efficient. This horizontally connected telecommunications network architecture could in the future actually be mirrored in terrestrial microcellular and picocellular systems as they tend to route calls more and more directly between end-users as the number of node connections tend to increase with more and more cells of smaller size being interconnected. One option in this respect would be in having a very-high-speed processor and switch installed on a High Altitude Long Endurance platform (at heights in excess of 18.5 kilometers) so that one megaswitch could connect hundreds of terrestrial cells in an urban microcellular system directly from one cell to any other, rather than traversing an hierarchical switching system.

It seems likely that fixed satellite services (FSS) will continue for a long time. Neverthe-less their role in the field of telecommunications will continue to change. In particular it seems most likely that FSS services will shrink as a percentage of all forms of satellite-based traffic. Certainly the role of fiber optic cable will likely expand within high-density and ur-ban regions as FSS shifts to rural and lower density routes or to broadcasting and network-ing roles. Clearly the role of FSS services as a trunk connector will erode as fiber claims more of this traffic. As on-board processing and signal regeneration allow for more direct routing of services between end-users, however, this could change. Business and private networks using increasingly low-cost VSAT and USAT networks are expanding. Figure 2.5 shows this trend, which represents the primary growth pattern for FSS services today. [31]

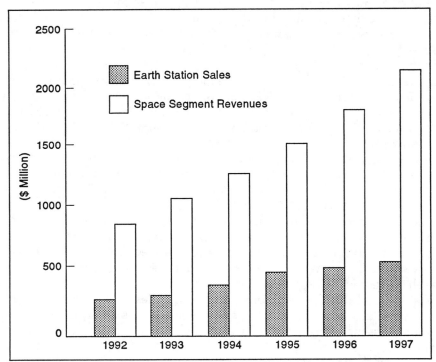

FIGURE 2.5   The Growth of VSAT and USAT Service in Private Business Networks: Revenue Forecasts by Segment (U.S.), 1992–1997
Source: Frost & Sullivan Market Intelligence

Satellite services are now much more than fixed or conventional FSS networks. The overall trend is for satellites to diversify to provide broadcast and television distribution services, mobile networking, navigational, plus rural and remote services. Indeed satellites will continue to serve as a key link to developing countries for decades to come in all ser-vice categories. Satellites will also still serve as a back-up to fiber optic cable for long-haul systems. Finally they will play a role in hybrid networks, especially if viable LEO and MEO systems can conquer the latency or delay problem. Unlike previous times, however, when growth rates for fixed satellite services often ranged from 15 to 30 percent per year, future growth may well remain under 8 percent per year.[32]

2.8.2 Mobile Satellites Providing Voice Service: Elliptical, Geosynchronous, Low and Medium Altitude Systems

The type of telecommunications where satellites enjoy a strong advantage is not surprisingly in the area of mobile communications. You do not plug cables into cars or trucks to provide mobile telecommunications for rather obvious reasons. Instead you use wireless radio communications. In many ways, one can look at a mobile satellite systems as a very tall tower from which a number of cells are provided and in which the Mobile Telephone Switching Offices (MTSOs) services are also conveniently available. Instead of covering a small, densely inhabited urban population, as in the case of cellular radio, the satellite cellular beams cover a much larger and likely much more unevenly populated area—sometimes in urban areas, sometimes in remote areas and sometimes over the ocean. The basic concepts involving terrestrial cellular telephone and mobile satellite service, especially low earth orbit satellite systems, are much the same. The only real difference is that the distances are much longer and the satellite platform is moving rather than stationary.

The options that are available for configuring a mobile satellite system are quite numerous. One option is to use an extremely powerful platform in geosynchronous orbit and blast down enough power to still reach very small transceivers. As one moves north or south of the equatorial plane toward higher latitudes, the geosynchronous approach becomes more problematic. This is because the look angle from the satellite to the earth's surface becomes lower and lower, making effective communications increasingly difficult. Also, transmission delays associated with an almost 72,000 kilometer round trip to geosynchronous orbit and back also constitutes another problem.

If you want a satellite system that can work to a truly compact, lightweight, and low-power transceiver on the ground with low delay and using a low gain, omnidirectional low-gain antenna, then there are rather limited options available. This solution demands a lot of rather low flying satellites in a tightly configured global constellation in order to accomplish both low delay and low path loss. The closer the satellite configuration is to the earth's surface, the better the performance and the better the frequency reuse opportunities, but it also means that more satellites must be added to the constellation to provide continuous coverage to support real-time voice service. This is obviously very expensive.

A medium earth orbit configuration, a compromise between low earth orbit and geosynchronous orbit, involves more delay and higher power levels to compensate for the greater path loss, but on the plus side, global coverage can be achieved with far fewer satellites. The proposed Odyssey mobile satellite system by TRW with 8000- to 10,000-kilometer orbits can provide global mobile voice services with some 18 satellites in contrast to the proposed Motorola Iridium mobile satellite system with 1000-kilometer orbits and a global constellation of 66 satellites.

There are other variations on the theme. Elliptical or Highly Elliptical Orbit satellite systems (EEO) and (HEO) can be designed to give attractive coverage to countries with higher southern or northern latitudes and the apogees of these orbits can be "stacked" over a particular country so as to provide continuous coverage for that area. The so-called Ellipso mobile satellite system plans to take this approach for mobile satellite services optimized for the United States. The global mobile satellite system, INMARSAT, the only fully operational mobile satellite system now capable of blanketing the earth, may ultimately use a combination of geosynchronous and medium earth orbit satellites if the so-called Project 21 or INMARSAT P is authorized to go forward. It will likely be a number of years before some clear-cut conclusions are reached as to which satellite orbits and operational approaches to voice-based, real-time mobile satellite services are preferred.[33]

Some envision that low earth orbit satellite systems will evolve to provide not only mobile services, but also a significant amount of rural and remote services that look a lot like fixed satellite services. The so-called "mega-LEO" satellite system, known as the Teledesic Satellite System with a proposed global constellation of 21 different planes and some 40 satellites in each plane, a total of 840 satellites, is one such system that has the band-width capability and network design capable of such a feat. This mammoth proposed satellite project with an estimated investment of some $10 billion (US) would have the capacity and technical characteristics to provide both mobile and fixed services.

The range of services proposed by the Teledesic system actually range from hand-held services at the kilobit per second level, to village-based systems operating at the megabit per second level, to even super computer interconnect at gigabit per second levels. Finally, some of the proposed "big LEO" satellite systems envision the use of intersatellite links to allow direct connectivity between adjacent satellites in space. This is a very new and expensive technology that also decreases the operating capacity of the satellite. As such, intersatellite technology is controversial with proponents strongly advocating the operational flexibility it provides and others suggesting it is an unnecessary, complex and expensive service feature that is not required.[34]

2.8.3 Store and Forward Mobile Satellite Communications Services—
"The Little LEOs"

The lower cost and somewhat less respectable cousins of the so-called "big LEO" satellite systems such as Motorola Iridium and Globalstar mobile satellite systems are the mobile satellite systems designed to support e-mail, data relay, fax, and other noncontinuous forms of telecommunications messaging. The idea is to create satellite systems that can provide low-cost messaging services that are fast, reliable, and totally global in coverage, but with the liability that they do not provide instantaneous and continuous connection between or among the end-users. These systems, simply by not offering voice service can suddenly become much simpler and much less costly to build and deploy. This is because the satellite configuration can now operate in low earth orbit with many fewer satellites. Typically, a configuration of eight to 20 satellites, depending upon the orbital characteristics and the volume of service carried, can provide effective global messaging service. These satellites can also be smaller and cheaper to build and launch.

The various satellite systems of this type can offer both mobile and fixed messaging service to very-low-cost terminals. These systems typically can relay a message from anywhere to anywhere on the planet in a few hours time and at a cost of about 1–100 bytes per 1 cent (U.S.). While these costs are high compared to conventional data relay services on developed telecommunications systems, they are very attractive when compared to the cost of very remote services or cellular systems. The biggest problem with such systems, often known as "little LEOs," is that of frequency allocations. The few frequencies available at 147 and 400 MHz, and even including the new allocation at 1 GHz, are still extremely limited in channel capacity and additional allocations will be needed for this service to significantly expand.[35]

Perhaps the most advanced and furthest developed of these systems is the ORBCOM system being manufactured and deployed by the Orbital Science Corporation. This system of 18 LEO satellites has been fully licensed by the FCC and is undergoing ITU coordination procedures. Launch of this system is anticipated in 1995/96. This system also includes GPS receivers and can carry out atmospheric measurements as well as relay data. Other systems such as the Russian Gonets, the University of Surrey microsatellite networks, and the rather sophisticated LEO One satellite system with intersatellite links are other systems in this service category. One ORBCOM will be offering commercial services in the immediate near term.

2.8.4 Direct Broadcast Satellite Services

This area, like mobile satellite services, is expected to expand rapidly in the next few years. Operational systems now are in service in Europe (ASTRA and BSB), Germany (TVSAT), France (TDF), and in Japan (BS-3 and soon N-Star). New DBS systems are also starting in the United States (DirecTV and Hubbard) and Mexico (Solidaridad). Quasi-DBS service is also available in North America through Primestar and ANIK satellites and in the Asia-Pacific region via INSAT and Asia SAT.

The concept of a Direct Broadcast Satellite, or in ITU terminology, Broadcast Satellite Service (BSS), is quite straightforward. The principle is one of broadcasting a signal from space with sufficient power to allow direct reception at a home or office with a very low cost and an inexpensive receiver that costs only a few hundred dollars (U.S.) to install. This is in contrast to satellite television distribution service whereby a signal is sent to a cable television system or over-the-air television station for redistribution to the home. Some systems such as ASTRA, however, have used FSS frequency allocations and high-powered antenna beams to provide a quasi-DBS service through satellites that are really technically Fixed Service rather than Direct Broadcast. These are technical distinctions that the consumer seldom appreciates. In time, as DBS becomes ever more powerful and digital processing also increases in performance, the distinction may become more clear cut. For instance, the highest-powered DBS systems can now provide a television signal to receivers that are only 35 centimeters or 14 inches in diameter—this is over ten times smaller than an ASTRA terminal for quasi-DBS service. The major appeal of DBS satellites in the United States has to date centered on the ability of digital compression techniques to all systems such as the Hughes DirecTV system to deliver 150 different channels to the home.[36]

The different classes of satellite systems, especially the most established geosynchronous satellite systems, tend to evolve from one set of characteristics into another.This progression can be seen by examining how fixed, then mobile and then broadcast systems become increasingly higher powered. To illustrate this progression in terms of these various "families" of satellite types, Figure 2.6 shows the move to higher power and thereby the shift to smaller, compact, and less expensive ground terminals.

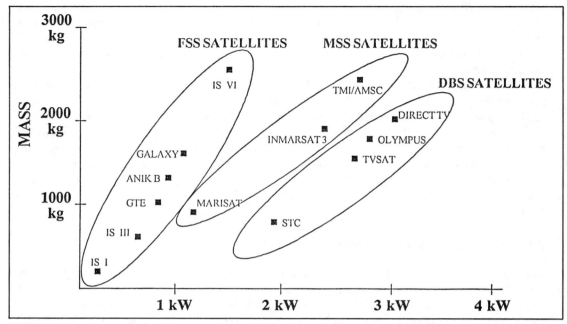

FIGURE 2.6   Comparison of Mass-to-Power Performance of Fixed, Mobile, and Broadcast Satellites

2.8.5 Military and Defense Satellite Communications Services

The military or defense applications of satellite communications are numerous. This is for a number of key reasons, which include the following: (a) the ability to provide mobile and tactical communications to a battlefield or remote, isolated locations; (b) the fast reaction ability to add or delete nodes in a network simply by installing earth station antennas; (c) the ability to provide a fully encrypted end-to-end communications link simply by providing a secure terminal at a new location. (This can be done today on a strictly commercial and civilian satellite system by having a specially equipped briefcase-sized INMARSAT M terminal located anywhere on the planet and at any land, air, or sea location). Satellites can be used for regular or nonstrategic military telecommunications purposes but in this case this would typically be in conjunction with or as back up to terrestrial telecommunications services. Satellites in this role today would often be used to provide redundancy or to reach isolated locations where fiber optic cable might not be available.

The military applications of satellites in the past have included mobile land, navy, and aircraft communications, earth-to-space and space-to-earth communications to support surveillance and remote sensing operations, radar systems, and nonstrategic operational and logistical communications. Military units in the United States, within NATO, within the former Soviet Union, and the former Comecon nations, as well as others have traditionally insisted on separate frequency allocations for these satellite services for a variety of reasons related to security, priority access and control, and simply because the frequencies were obtainable. In recent years, however, there has been increasing acceptance of the idea of providing certain military communications on commercial systems. The idea is simply that this is a more cost-efficient way of providing service and also a means of providing multiple path redundancy. Since commercial systems are not custom designed so as to benefit from economies of scale, they seldom include special design features such as radiation hardening or triple redundancy.

With larger-scale production and marketing, the cost benefits of commercial systems are often quite clear cut. In one instance involving the so-called Marisat System, the U.S. Navy actually operated one-half of the satellite for naval communications while the other half was operated on behalf of the INMARSAT maritime satellite system. It is estimated that this approach was three to four times less expensive than had the Navy designed a customized, highly redundant and "hardened" satellite themselves. Because of the specialized needs of military communications for flexible, mobile, and redeployable communications, it is believed that military communications satellites will be a continuing market even if more traffic is transferred to commercial systems where and when possible.[37]

2.8.6   Other Satellite Services

If one thinks of a satellite as being a very tall radio relay system, it is easy to recognize that satellites can be used for an amazingly broad range of telecommunications services. Anything that requires broadcasting services over a large region would make sense whether for television, HDTV, audio/radio, imaging, or data. Large networks, particularly asynchronous and asymmetrical ones, would particularly qualify for satellite services. Data relay from one satellite system to another such as from LEO to geosynchronous satellites to earth-based processing centers are another key application today as seen in the NASA Tracking and Data Relay Satellite System. Every satellite in orbit today whether for research, scientific data collection, remote sensing, meteorology, or some other form of application or exploration is using one or more frequencies to relay signals to earth.[38]

In space, atmospheric interference is no longer a problem and so it becomes possible to relay signals between satellites not just by radio waves but also by laser modulated signals. In the future, therefore, intersatellite links operating by means of optical telescopes and light modulation can be expected. In fact the Japanese experimental satellite, COMETS, will test these concepts with an operational prototype when it is launched in 1997. The Jet Propulsion Laboratory also plans to carry out experiments with interplanetary communications links using wireless "optical communications." In short, satellites are a versatile form of wireless communications that are made unique only because of the high orbital location and the particular satellite's movement. With these two differences accounted for, the satellite's payload is often much like the radio equipment placed on a tower or a mountaintop such as is the case with "normal" wireless communications.[39]

2.9   VHF, UHF, MICROWAVE, AND MILLIMETER WAVE
TRANSMISSION SYSTEMS

Many people today think of wireless technology as the means to provide mobile services to a car, truck, bus, or emergency vehicle. This is because such services are highly visible and millions of cellular subscribers, taxi cab customers, and even viewers of television shows about police and rescue work see this technology at work on a daily basis. It was not always thus.

At the turn of the century Guglielmo Marconi undertook a series of experiments to attempt to demonstrate long-distance wireless communications. He sent radio modulated signals up into the sky and then attempted to receive them at a distance by ships at sea. To the amazement of all, including Marconi himself, faint signals seemingly bent back from the heavens could be deciphered in code and then in voice. Marconi, like Alexander Graham Bell and Samuel F. B. Morse before him, was not a trained scientist, but rather an inspired tinkerer. No one at the time could explain why the signal was so providentially relayed back from the sky so that the message could be heard. Today we know in some detail about the characteristics of the "reflective" surface of the ionosphere. We also know that sunspot activity and other electromagnetic phenomena can break up this "electronic mirror" above the atmosphere and that shortwave communications is thus not an extremely reliable form of information relay, even with corrective multifrequency techniques.[40]

In subsequent years, knowledge accumulated about the characteristics of radio frequencies as they moved up through the electromagnetic spectrum. Different wavelengths were found to be best suited to different tasks. The demand for more and more services and broader band capacities constantly pushed the demand for spectrum. Although there were difficulties of precipitation attenuation and atmospheric scintillations with shorter wavelengths and with corresponding higher frequencies, these bands also afforded much broader spectra for useful allocations. The transmission requirements of television signals and broad band data in terms of broad band-width thus forced the movement upward. A single television channel typically requires 6 MHz of spectra. Clearly this requires a frequency assignment that is in the upper megahertz range or above. Otherwise it simply does not physically "fit" into the available spectrum. Today over-the-air television channels are assigned within the Very High Frequencies (VHF) or the 30–300 MHz band or the Ultra High Frequencies (UHF) or the 300–3000 MHz band. Even these assignments are today often seen as inappropriate, given the competing demand for these frequencies for say cellular radio and personal communications services.[41]

Critics, for instance, would say that television should be largely distributed over fiber optic or coaxial cable in order not to "consume" the scarce RF spectrum. In the case of mobile communications where cable connection is not feasible this is, however, not an option. The further thought is that if television is to be distributed by over-the-air broadcast it should be at much higher frequencies, such as millimeter wave bands, where large chunks of spectrum could much more easily be made available. This is, in fact, reflected in the experimental testing of the 28 GHz band for such purposes in such cases as the US WEST demonstrations in San Diego, California. Operational frequencies have still not been granted and satellite system users argue that they should have first claim on this band.

For the time being, however, most television for commercial entertainment, for public broadcasting, and for specialized educational purposes such as Instructional Fixed Television Service (IFTS) are send out on VHF or UHF band via high-powered transmitters from the top of very tall towers or the top of adjacent mountains. The same is true of AM and FM radio stations. In very large service areas, there may be repeaters or translator systems to extend the range of coverage or to serve areas that are blocked from receiving the initial signal.

Above the VHF and UHF bands come the microwave band that extends from 3000 to 30,000 MHz, or more simply 30 GHz. Since the gigahertz (GHz) band represents billions of cycles per second, the phrase for billions or "giga" is typically used. Even higher still is the millimeter wave band, so named because the wavelength is measured in millimeter lengths. This band begins at 30 GHz and continues up to 300 GHz. There remain radio frequencies even above this range, but there is very limited commercial or practical use for these bands at present. Telecommunications use as a practical matter thus jumps from millimeter wave all the way to the infrared and lightwave bands, where truly enormous bandwidths are available. Since there are problems of atmospheric or environmental interruptions when sending light or infrared signals through free space (i.e., wireless), the normal practice is to use such transmissions for very short distances such as a television remote control or wireless LAN connection in an office. Otherwise light or infrared transmissions are sent over fiber optic cables.[42]

The primary wireless application of microwave frequencies today is for satellite relay, which has been previously discussed, and for long-distance terrestrial microwave relay. Many of the older microwave systems use analog transmission, but newer installations are digital systems that can provide very-high-quality service. This is a means of relaying signals such a voice, data, and television from one location to another. Unlike television broadcast where signals are irradiated by omniantennas over the entire surrounding area, microwave relays use highly directional antennas to send messages from one relay station to another along a long-distance chain with towers spaced about 30–60 miles apart. This spacing is necessary due to the curvature of the earth, which requires a receiving tower to capture the signal before it literally goes off into space. The higher the towers or mountaintop, the greater the spacing between the relay stations. Finally in some remote areas or islands well offshore, there are longer distance relays called tropo-scatter, which are used to bounce signals to their locations. This is much like a shorter-range shortwave radio relay system. These tropo-scatter systems are largely obsolete but are still used in some isolated parts of the world.

Today most new broad band terrestrial telecommunications systems would tend to install fiber optic cable, but microwave still makes economic sense for many applications such as communications across mountains, swamps, bodies of water, or geological faults, as well as for spur connections to rural locations. Some carriers, who note that their networks are all fiber still "lease" microwave connections in remote areas.[43]

2.10   OTHER WIRELESS TELECOMMUNICATIONS

Ever since the experiments of Marconi nearly a century ago, new and improved ways to use radio waves for wireless communications have been found. In fact, they have grown, prospered, diversified, and become increasingly innovative. Unexpected uses and applications have been found in almost astonishing areas. One of the most unexpected was in the area of radio astronomy. Microwave antennas accidentally pointed to the skies registered faint but perceptible signals. After ruling out messages from extraterrestrials, scientists concluded that these mysterious radio sources were from stars and odd new phenomena known as quasars. These radio signals in many cases came from sources other than stars, or at least stars that optical telescopes could see. In time, scientists came to use optical, radio, X-ray and gamma ray telescopes to chart the heavens.

Today, a good deal of what we know about our galaxy, quasars, black holes, and even the layered structure of the universe stems from our accidental discovery of the art of wireless radio astronomy. Elaborate devices whose sensitivity can detect power levels 1000 times less than that represented by one snowflake falling to earth are now pointed to the sky at Jodrell Bank in England or Arecibo in Puerto Rico. The most sensitive device now at work in radio astronomy is in New Mexico, and it is engaged in the Search for Extra-Terrestrial Intelligence (SETI). This device is, in fact, dozens of highly sensitive radio astronomy dishes formed into a patterned array with each antenna working together through a central processor. This processor, with the help of artificial intelligence, integrates the received signal together to achieve super-refined and clear signals from the edge of the universe. This same basic concept of integrating multiple components together electronically are actually at work in the new phased array antennas being developed for mobile communications and direct broadcast satellite services.

This ability to communicate across thousands of light years represents the current limits of human-based radio communications. Considering the narrow limits of radio communications a century ago this is a remarkable achievement.[44]

There have been other spinoffs and extensions over the years. We have been able to establish links between scientific satellites as far away as Neptune. The three-axis-stabilized satellite design now used in commercial space systems began with space platforms designed for exploration of the solar system. Antenna designs perfected for radar tracking and then for radio astronomy have found useful applications in telecommunications with today's phased array antennas, as noted previously.[45] There is reason to believe that developments in superconductivity will, within the decade, find a host of applications for efficient telecommunications services. Some also believe that microwave-stabilized communications platforms will provide a new type of platform for broadcast and mobile communications. This very new technology known as high altitude long endurance (HALE) platforms can in theory be deployed more cheaply than satellites and also be brought back to the earth's surface for repairs, retrofits, and capacity upgrades.[46]

The rapid cycle of innovation in telecommunications today gives no indication of diminishing with time. Today there are literally thousands of uses for telecommunications, and these new services and applications also continue to expand. The possibilities seem almost limitless. They include voice, data, fax, e-mail, television, scientific visualization, interactive CAD/CAM, multi-media, virtual reality, electronic games, high-definition imaging, HDTV, radio, and audio. They also include services that will decline over time such as long-distance microwave relay and tropo-scatter systems. In addition, they include services that may well never see widespread deployment such as millimeter wave-guides. They include satellite services of all types including those for broadcasting, telecommunications relay, navigation, tracking, intersatellite relay and satellite data relay for application and scientific missions. Certainly they include all forms of mobile services, which will increasingly duplicate the fixed terrestrial telecommunications services including mobile video and imaging services.

In the business environment there are a host of new capabilities and possibilities as well. These include wireless PABXs and wireless and even satellite-based LANs. Within a decade radio and infrared bus technology will likely provide virtually any type of office telecommunications service.

In general, wireless technology offers the flexibility to interconnect mobile users or locations that need on occasion to be reconfigured or even redeployed. Wireless communications services are certainly not a panacea, but they are often cost efficient, easily accessible from remote locations, and highly mobile. If there is a concern with wireless technology it is often associated with atmospheric interference, multipath, back-scatter, blockage and ghosting from tall buildings, vegetation, mountains, or low look angles. There can be problems with lower-powered systems involving large coverage areas. In the field of satellites there are concerns with path loss and even transmission delays—particularly where geosynchronous satellite systems are involved. In the field of telecommunications there is no single technology suitable for all services, applications, or locations. The best overall design for a global telecommunications system is thus a hybrid network built up of different systems optimized for specific purposes, but integrated by means of universal open standards that allow all systems to interconnect seamlessly and at low cost.

ENDNOTES

(1) Berger, J., "How Do I Get There? Navigation in the "Smart Car" Age," *Wireless,* May/June 1994, Vol. 3, No. 3, pp. 28ff.

(2) Nakonecznyj, I.T., "The Wireless Revolution—It's Here Almost!," The NEC ComForum, Orlando, FL, 1992; also see "Amendment of the Commmission's Rules to Establish New Personal Communications Services," GEN Docket No. 90-314, PCC 93-451, adopted September 23, 1993 and released October 22, 1993, pp. 1939ff.

(3) "Enhanced paging with digital messaging," *Wireless,* December 1993, Vol. 2, No. 4; also see "Motorola's Cellular Phone-Page," *Wireless,* September 1993, Vol. 2. No. 3.

(4) The RF Spectrum, ITU Region 2, (Denver, CO: Communications Technologies, Inc., 1993).

(5) Hoffman, R., "DGPS: Will DOD Lower Its Guard?" *Satellite Communications,* April 1994, p. 46; also see "Fleet Track To Feature Enhanced Digital GPS Capability," *Global Positioning and Navigation News,* March 10, 1994, Vol. 4, No. 5, p. 8.

(6) "The Case for Multi-Purpose Frequency Allocations," *Satcom Quarterly,* June 1992.

(7) Young, W.H., "AMPS: Introduction, Background and Objectives," *The Bell System Technical Journal,* January 1979, pp. 1-14.

(8) Berger, J., "SMR Shows Its Utility," *Wireless,* September 1993, Vol 2, No. 3, pp. 34-35.

(9) RF Spectrum, ITU Region 2 (Denver, CO: Communications Technologies, Inc., 1993).

(10) Berger, J., "SMR Shows Its Utility," *Wireless,* September 1993, p. 35.

(11) MacDonald, V.H., "Advanced Mobile Phone Service: The Cellular Concept," *The Bell System Technical Journal,* January 1979, pp. 16-22.

(12) Young, W.H., "AMPS: Introduction, Background and Objectives," *The Bell System Technical Journal,* January 1979, pp. 1-12.

(13) NASA/NSF, *Panel Report on Satellite Communications Systems and Technology* (Baltimore, MD: International Technology Research Institute, July 1993); see sections on power systems and on site visit to Toshiba.

(14) Hoffmeyer, J.A., "Personal Communications Services," *ITS Technical Report* (Boulder, CO: Institute for Telecommunications Sciences, 1994), pp. 1-24.

(15) Calhoun, G., *Digital Cellular Radio* (Norwood, MA: Artech House, 1987), pp. 98ff.

(16) "Interstate Highway Based Cellular Systems," *Wireless*, January 1993, Vol. 1, No. 1.

(17) Manuta, L., "Big Leo Equals Big Deal," *Satellite Communications*, April 1994, p.14; also see Steele, L.C., "The 1992 World Administrative Radio Conference: New Allocations and New Challenges for Satellite Communications," July 1992, pp. 19-24.

(18) Kirlin, K., "SIM Technology Promotes Global Roaming and UPT Services," Boulder, CO, University of Colorado, 1994.

(19) Kachmar, M., "The Virtual Office: Mobile Professionals Reach for Total On-the-Road Functionality," *Wireless*, September 1993, Vol. 2, No. 3, pp. 24-27.

(20) Brown, J., "Data over Cellular," *ICA Expo*, May 1994, Conference Proceedings; Virtual Office schematic from *Wireless*, Fig 2.5, Vol. 3, No. 3, pp. 12-14.

(21) Nakoneczny, I.T., "The Wireless Revolution—It's Here Almost!" The NEC ConForum, Orlando, FL, 1992, pp. 1939-40.

(22) "Ameritech Ices Digital Cellular," *Wireless*, September 1993, Vol. 2, No. 3, p. 7.

(23) Hoffmeyer, J.A., "Personal Communications Services," *ITP Report* (Boulder, CO: Institute for Telecommunications Sciences, 1994), pp. 1-20.

(24) "PSC Services," Northern Telecom Presentation, Richardson, TX, April 1993.

(25) Hoffmeyer, J.A., "Personal Communications Services," *ITP Report* (Boulder, CO: Institute for Telecommunications Sciences, 1994), pp. 1-20.

(26) "Wireless LANS/PBX," *Wireless*, December 1992, Vol. 2, No. 4.

(27) McCarthy, W., "Chase Manhattan Installs Infrared Token Ring," *Wireless*, September 1993, Vol. 2, No. 3, pp. 49-50.

(28) ITU Radio Regulations, "Definition of Space Communications Services" (Geneva, Switzerland: ITU, 1992); also see "Domestic Services on INTELSAT System," (Washington, DC: INTELSAT, 1993).

(29) Pelton, J.N., "The Next 100 Years in Satellite Communications," *Via Satellite*, September 1992.

(30) Carraway, R.L., J.M. Cummins, and J. R. Freeland, "The Relative Efficiency of Satellites and Fiber-Optic Cables in Multipoint Networks," *Journal of Space Communications* (Amsterdam, Netherlands: IOS Press, January 1989, Vol. 6, No. 4, pp. 277-289.

(31) Bull, S., "International Trend in VSATs," *Via Satellite*, December 1994, pp. 30-32; "VSAT Systems Growth," *Satellite Communications*, April 1994, pp. 12-15.

(32) Frieden, R., "Wireline vs. Wireless: Can Network Parity Be Reached?" *Satellite Communications*, July 1994, pp. 20-23.

(33) *Proceedings of the International Mobile Satellite Conference* (Pasadena, CA: JPL, May 1993).

(34) Pelton, J.N., "Will the Small Satellite Market Be Large?" *Satellite Communications*, March and April 1992.

(35) ———, "Will the Small Satellite Market Be Large?" *Satellite Communications*, March and April 1992.

(36) Marshall, P., "Global Television by Satellite," *The Journal of Space Communications and Broadcasting*, January 1989, Vol. 6, No. 4; also see other articles in this issue.

(37) "The Future of Military Communications and Its RF Spectrum Requirements," Rand Corporation, Santa Monica, CA, and Washington, DC, 1990, pp. 20-35.

(38) NASA/NSF, *Panel Report on Satellite Communications Systems and Technology* (Baltimore, MD: International Technology Research Institute, July 1993); see sections on power systems and on site visit to Toshiba.

(39) NASA/NSF, *Panel Report on Satellite Communications Systems and Technology* (Baltimore, MD: International Technology Research Institute, July 1993); see sections on power systems and on site visit to Toshiba.

(40) Calhoun, G., *Digital Cellular Radio* (Norwood, MA: Artech House, 1987), see Chapter 2.

(41) Manuta, L., "Riding the Spectrum Wave," *Satellite Communications*, July 1994, pp. 24-26.

(42) "Exploiting the SHF and EHF frequency bands," *Satcom Quarterly*, Jet Propulsion Lab., September 1993.

(43) "Fundamental of Radio Communications" (Boulder, CO: U.S. Department of Commerce, Institute for Telecommunications Science, U.S. Government Printing Office, 1990).

(44) "The Case for Exclusive Frequency Allocations for Radio Astronomy," *Astronomer*, January 1992, pp. 14-18.

(45) "The Future of Military Communications and Its RF Spectrum Requirements," Rand Corporation, Santa Monica, CA, and Washington, DC, 1990, pp. 20-35.

(46) Yam, P., "Current Events: Trends in Superconductivity," *Scientific American*, December 1993, pp. 119-126.

PART II

THE MARKET, THE SERVICES, AND THE TECHNOLOGY

CHAPTER THREE

THE WIRELESS AND SATELLITE MARKETS

3.1 EXAMINING THE WIRELESS MARKET

The previous chapters discussed the basic concepts in wireless telecommunications, the services and applications and, in broad terms, the rapid growth potential of the entire wireless market both within the United States and globally. It seems indeed that the emergence of new technology and the burgeoning of new services and innovative applications will continue to expand well into the next century. This chapter provides a deeper probe into the various components that will stimulate and define the future combined market for wireless communications.

In the 1950s, 1960s, and 1970s entertainment, business communications, and education- and health-related services were all largely thought of as "static" or "fixed" location activities. These services were typically delivered to a single location either in the home, the school, the hospital, or the business office. More specifically these activities were thought to occur at desks or in front of living room television sets or other fixed locations, such as one or two home or business telephones. Beginning in the 1980s and in the 1990s, however, it has seemed natural and even expected in the United States and many other OECD countries to receive all forms of information "on the go." Those raised in a telecommunications milieu characterized by the walking CDs, car phones, mobile faxes, electronic tablets, and "palm-top" computers, not surprisingly, tend to see this move toward increased mobility as being quite normal. It is hard to escape the notion of complete mobility in an age where it is now possible to watch a television image projected onto dark eyeglasses. The concept of handless and cordless communications is spreading throughout society in the United States and many other countries around the globe as well.[1]

The question is thus not whether there will be market growth, but rather how much and in which sectors. There are, of course, many methods for forecasting future markets. These include the following: Delphi surveys or focus group discussions among market experts; extrapolations of past trends; low, medium, and high range projections; parallel trends analysis; market scenarios; heuristic models; decision trees coupled to benefit/cost analysis of key options; simulations; and so on. All of these techniques have their advantages and drawbacks, and none are guaranteed to give good or even reasonably accurate results.

There is at least one approach that seems to make a great deal of common sense in terms of obtaining general background information and in establishing key market trends. This, quite simply, is what might be called a detailed environmental assessment of the wireless telecommunications market. In strategic planning an environmental assessment is the first step in trying to carry out an evaluation of strengths, weaknesses, opportunities, and threats—or more simply a SWOT analysis. This chapter therefore presents an environmental assessment for all major wireless services and applications in terms of the past, the present, and the projected future market.

3.2 THE ENVIRONMENTAL ASSESSMENT

The continued growth and expansion of wireless services seems likely to continue across a wide spectrum. There are growing wireless markets for education and training activities, for health and medical services; for business, financial, and information activities; and for energy and transportation applications. The most complicated area is that of entertainment and leisure services because one currently sees growth and shrinkage occurring simultaneously such as growth in DBS services, but a market flattening or even shrinkage of over-the-air television broadcasting. Overall expenditures for entertainment and advertising is in national market surveys showing only modest growth consistent with inflation and the expansion of the GNP.

Current trends also suggest ongoing growth associated with police work, criminal justice, human services, education, health care, and economic and social development. Less clear, however, is what might be the patterns related to military and peace-keeping operations, retailing and merchandising, agriculture, manufacturing, and mining. These areas may well exhibit more modest growth through the 1990s. It seems significant that those sectors of both the United States and the global economy that are major users of information and telecommunications and that are oriented toward world trade and toward productivity gains are also the areas where perhaps the greatest growth can be anticipated regardless of whether wire or wireless services are concerned.

In the past several decades the countries of the OECD have moved toward service economies that have served to fuel all forms of telecommunications growth, including wireless. Meanwhile telecommunications growth within the "so-called" developing or industrializing countries of the world, supported in part by development grants and loans, has been even more rapid. Overall these gains in absolute terms have still closed the gap in a significant way for only the most advanced of the industrializing countries such as Singapore, Taiwan, Hong Kong, and South Korea. Furthermore, these gains by developing countries have typically been in wire and cable systems as opposed to wireless. Again the so-called "dragons" and "tigers" of Southeast Asia are the exception to the rule. In the past decade overall telecommunications development grew about 6 percent per year in North America, 7 percent in Europe and Japan, and about 9 percent in the rest of the world. This statistic is misleading in that some of the industrializing countries grew at rates of 15 to 20 percent per annum while other developing countries with less resources were growing much more slowly. As of the end of 1994 all telecommunications services represent about $550 billion in annual revenues, and this level will likely rise to $1 trillion by the year 2001. These global market trends for the overall field of telecommunications services are seen in Figure 3.1.[2]

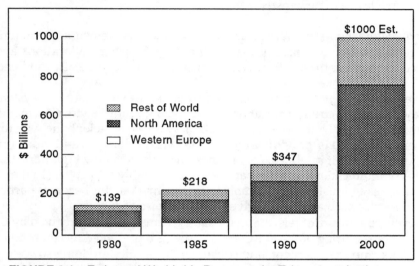

FIGURE 3.1 Estimated Worldwide Revenues for Telecommunications

Whether wire or wireless technology will drive the future of telecommunications has been the subject of a great deal of analysis and comment. In particular, at least two major viewpoints on what may happen over the next 20 to 30 years are provided in Figures 1.3 and 1.4. These alternative views have been called the Negroponte Flip and the Pelton Merge, respectively. As noted in Chapter 1, the viewpoint provided in the "so-called" Flip is that of a sudden and rather radical shift of usage patterns with most current wireline service for voice and data links to the home and office being replaced by cellular and PCS connections within the next 15 years. It also predicts that most broadcasting services will switch during this same time-frame from over-the-air broadcast to cable television or at least its twenty-first century equivalent. In many ways it suggests a major win for fiber optic cable, a major loss of satellites and over-the-air broadcast, and major gains for the new ESMR, PCS, and digital cellular services.

The Pelton Merge suggests, on the other hand, a less dramatic shift in the technology. This in turn suggests a more gradual change in customer-service profiles, in actual consumer and business applications, or in frequency reallocations. Thus, this more "conservative" scenario predicts that there will be "wins" and "losses" for both wire and wireless technology, but that the real winner may well be "hybrid" wire and wireless systems that are integrated together. While some shifts can and will occur, the Pelton Merge suggests that "hybridization" will be a more likely result than a complete and tumultuous 15-year flip-flop. The full-scale Negroponte Flip is not predicted to occur for the following reasons.

Economics.   The radical flip-flop would require hundreds of billions if not trillions of dollars to be invested. Neither business people nor regulators can reasonably agree to such rapid and massive level of investments. Further studies by advertising firms and consumer buying trends have not identified products or services with sufficient new value added to explain massive levels of new investment.

Sunk Investment and the Comfortable Status Quo.   A tremendous amount of good and serviceable equipment, facilities, as well as key frequency allocations and licenses would need to be prematurely abandoned, transferred, or substantially modified.

Obsolete Technology Has a Way of Coming Back.   Experimental systems have shown ways to send up to a 100 megabits per second over copper wire via a method known as Asymmetrical Digital Subscriber Line (ADSL). The "passé" telephone wire to the home may still prove capable of providing multiple television channels and video on demand. The new low earth orbit satellites, for instance, may create a new high-performance low-latency telecommunications service that proves to be highly attractive. (To the extent telephone companies seriously enter the cable television market, this becomes particularly relevant.) DBS service such as DirecTV gives great promise of success.

In any event, the future of telecommunications seems likely to be driven by the merger, overlap, and restructuring of the five C's industries as represented by communications, cable television, computers, content, and consumer electronics. In doing so we may, in particular, anticipate that telephone companies, cable television industries, and broadcasters may start to merge. There is much more involved here than the now defunct Bell Atlantic and TCI merger or the very much alive US WEST and Time-Warner marriage. It is also AT&T and McCaw, British Telecommunications and MCI, and the Paramount megadeal. The chart provided in Figure 3.2 as prepared by Goldman-Sachs suggests that all of these industrial groups may seek partnerships, mergers, or acquisition.[3]

A breakdown of key services in the wireless telecommunications field is provided in Figures 3.4 and 3.5. This chart combines the estimates of some 20 experts for both the United States and the OECD countries. It does not cover the developing countries since the trend lines, particularly for wireless services, are not necessarily parallel. In the developing world today there is, in fact, rapid growth in telecommunications, but this does not yet translate into rapid growth of wireless systems. In the developing societies during the 1980s and early 1990s, there was limited investment in terrestrial wireless service. Radio-based services in these countries were largely focused on domestic, regional, and international satellite and some radio and television broadcasting systems.

Global trends for overall telecommunications and information technology development are well established and documented by a wide range of statistical data. The likely increase of all forms of telecommunications and information technology is really not in doubt. In general, telecommunications has expanded more than twice as fast as the overall global economy, while wireless services have expanded more than three times as fast, especially

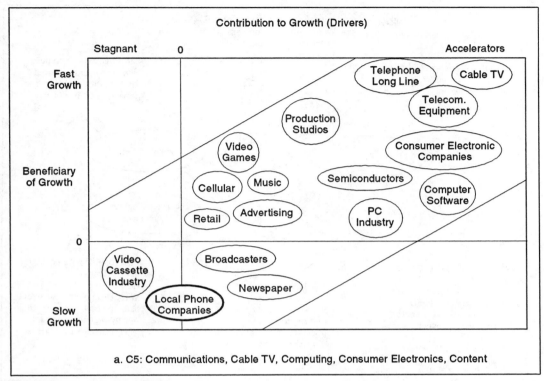

FIGURE 3.2   C-5 Convergence

within OECD countries. The more difficult issue is thus whether "wire" or "wireless" technology will "win out" over the other. Will one of these basic types of technology or some important new applications become predominant in the telecommunications field? Alternatively, will some form of dynamic balance be achieved so that rapid parallel growth somehow is achieved? Finally there are questions as to whether certain specific services and applications will become predominantly wire or wireless in the future.

The overall field of telecommunications and especially mobile telecommunications is thus beset by many forces of change both from within and without . Today the primary forces of change can be thought of as being driven by at least six key factors. These are: (a) convergence, (b) globalization, (c) technological advances, (d) new forms of competition, (e) changing demand patterns and markets, and (f) deregulation and liberalization. These forces and their uncertain impact on the field of wireless telecommunications are shown in Figure 3.3.

As this process of convergence occurs, then the "Pelton Merge" of "wire" and "wireless" technology into hybrid systems could well be expected to accelerate. The survey of experts described earlier in Chapter 1 also solicited information on net market gains and/or losses. This survey was conducted on the basis of net household and offices serviced in order to show real growth (or loss) independent of inflation. It shows for the most part some major increases. What is perhaps most significant is that these major and perhaps optimistic

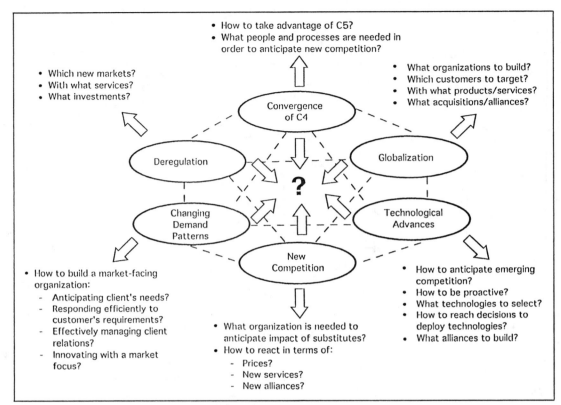

- How to take advantage of C5?
- What people and processes are needed in order to anticipate new competition?

- Which new markets?
- With what services?
- What investments?

- What organizations to build?
- Which customers to target?
- With what products/services?
- What acquisitions/alliances?

Convergence of C4

Deregulation

Globalization

?

Changing Demand Patterns

Technological Advances

New Competition

- How to build a market-facing organization:
  - Anticipating client's needs?
  - Responding efficiently to customer's requirements?
  - Effectively managing client relations?
  - Innovating with a market focus?

- What organization is needed to anticipate impact of substitutes?
- How to react in terms of:
  - Prices?
  - New services?
  - New alliances?

- How to anticipate emerging competition?
- How to be proactive?
- What technologies to select?
- How to reach decisions to deploy technologies?
- What alliances to build?

FIGURE 3.3   Six Forces of Change in Telecommunications

changes still do not reflect the idea of a sudden and dramatic "flip-flop" in terms of the overall structure and revenue profile of the telecommunications industry. This observation applies to both the United States results as well as the projections for the countries of the OECD as well. These projections as to future wireless services and applications are provided in Figures 3.4 and 3.5.

An additional key observation is that the largest increase, namely in the area of PCS, must also be considered the most uncertain. This is because it is in this area where the most unknowns also now exist. This includes a lack of clarity with regard to standards, nature of the service, market demand, technical solutions, and charging structure.

These projections, as shown in Figures 3.4. and 3.5, indicate rapid growth of most segments of the wireless market. This suggests there is little in the hybridization trend to slow the spread of wire, cable, or free-space communications. In fact this may result in mutual stimuli as they facilitate the needs and demands of a diverse customer base.

This projection for the U.S. and the OECD telecommunications market as reflected in Figure 3.4 and 3.5 contains an implicit assumption that there will indeed be major shifts in services. Even so, this forecast is very much within the context of current growth trends for

Service Description	Net New Households Change		% Annual Growth	
	Low	High	Low	High
1. DBS Service	7 M	20 M	12%	25%
2. PCS/Cellular (Subscribers)	17 M	32 M	18%	30%
3. Wire-based Telephone	−4 M	3 M	−2%	1.5%
4. Cable Television/Wireless Cable	5 M	7 M	3%	5%
5. Integrated Cable TV/Telephone Service	4 M	5 M	7%	10%
6. UHF/VHF TV Over-the-Air	−15M	−25 M	−12%	−20%
7. Mobile Cellular Satellite Service (1997–2000 only)	2 M	3 M	10%	13%
8. Wireless LANs/Wireless PABXs	2 M (Offices)	3 M (Offices)	8%	12%

Note: To reduce the impact of different tariffing concepts, inflation, and so on, these figures were developed on the basis of user numbers rather than revenues or dollars.

FIGURE 3.4 High and Low Forecasts—Basic Shifts in the U.S. Telecommunications Service Mix (Net Estimated Gains or Losses by Household 1994-2000)

Service Description	Net New Households Change		% Annual Growth	
	Low	High	Low	High
1. DBS Service	15 M	25 M	10%	15%
2. PCS/Cellular (Subscribers)	35 M	45 M	15%	21%
3. Wire-based Telephone	8 M	12 M	2%	3%
4. Cable Television/Wireless Cable	11 M	16 M	4%	7%
5. Integrated Cable TV/Telephone Service	9 M	15 M	10%	15%
6. UHF/VHF TV Over-the-Air	−20 M	−32 M	−10%	−12%
7. Mobile Cellular Satellite Service (1997–2000 only)	4 M	6 M	10%	13%
8. Wireless LANs/Wireless PABXs	3 M (Offices)	4M (Offices)	7%	10%

**The countries of the OECD include the United States, Canada, Japan, Australia, New Zealand, and all of Western Europe. Again the number of users rather than revenues were used to develop these projections on the basis of an absolute common denominator.

FIGURE 3.5 High and Low Forecasts—Basic Shifts in OECD** Telecommunications Service Mix (Net Estimated Gains or Losses by Household Through 2000)

wire and wireless technology. In terms of shifts in frequency allocations this could well mean television and radio broadcasting services being shifted to personal communications services. It also assumes wire telephone services will begin transferring to cable television connections to the home. Even so this is still what might be called a "modified status quo" projection. Based on the above projections it is still difficult to project within the various constraints noted above much more than a 25 percent shift from today's mix of services within the next decade as opposed to a complete or nearly total flip-flop. Studies conducted by the Center for Telecommunications Management (CTM) and the International Engineering Consortium (IEC) of telecommunications market trends have generally shown similar results in terms of growth rates and swings in wire- and wireless-related service demands.

It should be noted that some trends will be in the opposite direction. Direct broadcast satellite services, if they succeed in capturing market share in rural and remote areas and in certain urban groups for both entertainment and professional training purposes, will serve to counteract the wireless to wire transformation. If telephone and entertainment services combine on cable television systems, this will also serve to reduce the wire to wireless switch from the opposite perspective. Thus, if the projection shown in Figures 3.4 and 3.5 turn out to be more or less on target, then the so-called "flip" will tend to look much more like a quarter cartwheel.

To seek to analyze the implications of this macrolevel scenario in greater depth, the projections shown in Figures 3.4 and 3.5 were developed on an application by application basis. After these figures were developed, high and low estimates for the overall marketplace were also derived by setting the high and low standard deviation from the median response over two rounds of surveys among U.S. and international experts.

Thus, the forecast figures in the above figures were developed in the first half of 1994 by compiling the survey results provided by 20 telecommunications experts drawn from carriers, vendors, and users. There is no particular "magic" in these predictions nor is there any reason to believe that actual growth or decline in the market will even be within the high or low limits that this survey produced. The more interesting result is that regardless of whether one accepts the high or low forecasts, the market shifts are in no way like the radical transformations that the Negroponte Flip might predict. These survey results thus seem to reflect what might be called a "reality filter" that factors in market, economic, regulatory, and societal conditions.

One key environmental factor, however, is clear and that is the desire for business and residential consumers to be more and more mobile, flexible, and free of time and place constraints. Just in the business environment detailed studies have been undertaken of who needs and wants complete freedom of mobile communications. One study projects that by the year 1997 the profile of corporate mobile communications users will break down as follows: (a) salespeople—26 percent; (b) field service personnel—23 percent; (c) administrators—18 percent; (d) field engineers—15 percent; (e) senior management—10 percent; and others—8 percent. This suggests that the profile of users will be more and more distributed and pushed more and more toward the top of the corporate pyramid.[4] This process of progressive development of mobile communications is presented in Figure 3.6. [5]

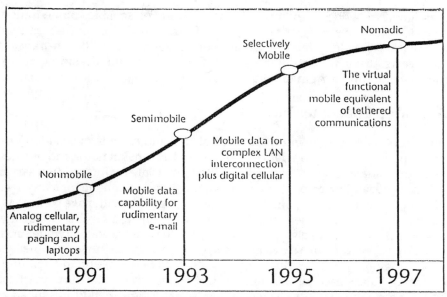

FIGURE 3.6 Stages of Mobile Professional Integration. *Source*: Courtesy of the Yankee Group.

3.3 MARKET FACTORS THAT WILL SERVE AS A BRAKE ON RAPID SHIFTS IN THE FIELD OF TELECOMMUNICATIONS

There are at least seven real world factors that would seem to inhibit or delay the longer term trend toward "wire to wireless" transformation and *vice versa*. These key factors as identified in the survey process are as follows.

3.3.1 Need for New Frequency Allocations for Digital Cellular and PCS

There are currently new frequencies being assigned to enhanced mobile communications. Likewise, improved frequency reuse techniques are being developed and better digital compression systems are being rapidly implemented. Nevertheless, currently available frequencies could at best support a fraction of the demand if digital cellular was to become the primary telecommunications means for all users in the United States or in the OECD countries. This is essentially true if PCS is implemented with larger cell sizes than the originally conceived microcells. This seems by far the most likely assumption.[6]

3.3.2 Vested Frequency Allocations for Radio, Television, Military Systems

No one with sunk investment, an established customer base, and wide spread user equipment in an established market readily surrenders it. The military establishment and over-the-air broadcasters who now control much of the most interesting spectrum will not give up their assigned frequencies without a fight and some reasonable compensation. There may well be longer-term strategies for migration from lower to higher frequencies or from wireless to wired technology, but this would likely involve ten- to 20-year transitions, not three to five years. Just in the case of the new 28-GHz services, there is strong opposition to the frequency reallocations from satellite users and others.[7]

3.3.3 Available Consumer Income

Some consumers may be able and willing to subscribe to cable television, buy premium television channels plus video on demand, purchase DBS, install home and business ISDN lines for voice and data services, buy value-added financial and information services, and obtain PCS services and a personal telephone number and telephone for every member of the family, but this will be a very small minority. Most people will make choices as to how much of their disposable income they will invest in information services. Most lower- and middle-income people will likely opt for basic flat-rate wire-based telephone service over more expensive wireless mobile services. (This trade-off will, of course, become more complicated if the regional Bell operating companies should be successful in obtaining regulatory approval of "measured" or per-minute charges for local calls.) Surveys of consumer spending on information services as well as net advertising expenditures by ad agencies show a relatively flat expenditure curve in the United States over the last decade.[8]

3.3.4 Consumer Acceptance Rates

With most new products and services, there is usually a period of experimentation with the adventurous testing the new offerings. This is often followed by the first of the mainstream consumers, and eventually a bridge is found between the innovators and the mass market. After a base of perhaps ten percent is reached, the true mass marketing profile comes into play. In the area of new telecommunications products and services, precursor markets are important. Analog cellular is a clearly useful antecedent to PCS and digital cellular in that consumer acceptance at a much higher rate can be projected. Nevertheless consumers are cautious and "overnight" acceptance of new products and services simply does not happen. In telecommunications services as in most market patterns a bell curve effect is usually seen.

3.3.5 Dynamic Tension Between Broadcasters and the Cable TV Industry

The notion that cable television will replace the traditional broadcasting corporations overlooks a key fact. The over-the-air television networks do more than broadcast signals through affiliate stations. They are still the most important and largest single source of programming for cable television. Under the 1992 Cable Act, cable television organizations were required to negotiate agreements with the broadcasters concerning the carriage of the networks programming. Since the real source of the U.S. television networks' income is advertising revenues, the networks and their affiliate stations settled for rather modest fees in order to maintain the highest levels of viewership via cable channels. In short, there remains an important symbiotic relationship between cable and the over-the-air networks.

This condition is, at least to some extent, also true in other parts of the world as well.[9]

The most important point is that the relationship between cable television and "conventional" television programmers is in a state of rapid flux. These key elements of change are as follows:

(a) The integrated cable television systems that both produce and distribute programming, such as Time-Warner/HBO, are producing more and more of their own programming.[10]

(b) The cost effectiveness and technical feasibility of international exchange and especially of transoceanic television programming has greatly increased. Today over 100 international satellite channels are now in operation via INTELSAT, Pan Am Sat, EUTELSAT, and other systems such as BSB, Asia Sat, and ASTRA.[11]

(c) The advent of new direct broadcast satellite systems. These include the Hughes DirecTV system, the Hubbard DBS system, the Japanese N-Star system, the follow-on DBS systems of ASTRA, British Sky Broadcasting (BSB), the German TV SAT, and the French TDF systems. These DBS satellite systems will both greatly enrich and complicate the delivery of television programming to the home. They will also impact those who produce television programming.[12]

(d) The increasing use of digitally compressed television delivery cable television systems, which also will allow the effective distribution of High Definition Television (HDTV), interactive television, and CD-quality radio. Digitally compressed signals will over the next decade make conventional, over-the-air analog television and radio systems technologically obsolete, cost ineffective, and especially, highly wasteful of available radio spectrum. This is not the case today, however, and from a business perspective a decade is a long time.[13]

The conclusions that one might reach from these trends are that "conventional" over-the-air television broadcasting is today not yet obsolete. Certainly, it is still a powerful source of programming. Even so, conventional broadcasting is clearly in a retrenchment mode. These trends suggest that television broadcasters in the future will control significantly less of the future distribution modes. The conventional over-the-air broadcasters are not well positioned to provide innovative new broad band and interactive video services. Broadcaster's large claims on the radio spectrum will be increasingly under attack, and over time reallocations can be anticipated, especially when billions of dollars of new governmental revenues can be derived from such redeployment of frequencies.

What might the "wireless" television world do to respond pro-actively to these trend lines? The answer is actually rather clear cut and predictable. This is clear, in part, because it is indeed starting to happen. The obvious adjustments are to transform as rapidly as possible to digital modulation and advanced encoding techniques to provide a better quality of service and with much less spectrum required per television or HDTV channel. One concept would allow broadcasters to use digital compression techniques to derive six channels from the same spectrum now used by a 6-MHz channel. This would allow conventional broadcasters to provide not only a greatly expanded range of programming but also pay-per-view and other enhanced value services.

Also, there are the so-called MDS or "wireless" cable television services operating in the microwave bands, which are creating competitive markets separate from the wire-based conventional cable television systems. These MDS services now have a customer base that numbers in the millions and are doing so by essentially "fighting fire with fire." Namely, they are using cable TV programming to compete with both conventional over-the-air broadcasters and conventional cable television systems. So there will also be the option of LMDS service, which will accomplish the same thing but at higher frequencies and with cellular technology so as to achieve 100-channel television systems. This LMDS technology could also add interactivity as an additional service feature. There is, however, opposition to frequency reassignments and "squandering" more spectrum on wireless services through the LMDS service.[14]

The third step, which is closely linked to the second step, is to utilize the multiple reuse techniques that have been successfully developed and implemented in the world of cellular radio telephone and transition television directly into telecommunications services. The application of such advanced cellular techniques would make the use of these new frequencies cost effective. There is little indication that the broadcasting industry is equipped to consider such a bold set of moves to become a technology leader in the highly competitive telecommunications industry.

The exact same logic that was first developed for mobile telephone services is now, in fact, being tested for cellular broadcasting cells in the United States and in a few other countries as well. If the broadcasters are not prepared to move aggressively in this area, then others are. US WEST is one of dozens of U.S. corporations that are experimenting with 28-GHz cellular television broadcasting in cities such as San Diego, California and Phoenix, Arizona. Also, in 1994 US WEST joined with Air Touch, the cellular spinoff of PacTel to create a strong new global entity that rivals McCaw Communications as the largest worldwide provider of cellular services. The Bell Atlantic and NYNEX mobile telecom alliance suggests that moves to achieve key strategic partnerships and new economies of scale can be expected to continue. Depending upon the cellular pattern of frequency reuse, this technology can expand the efficiency with which a frequency is used by five, six or even ten times. There is a good deal of logic in using cellular patterns for broadcasting applications. It is also clear from the preexisting use of cellular radio telephone at lower frequencies that one could use the millimeter wave bands for both mobile communications as well as for broadcasting purposes, even though some separation of the two applications would be necessary.[15]

To an extent, these new wireless applications suggest a continuing symbiosis of wire and wireless technologies and thereby "hybrid" markets. The advent of DBS television distribution, millimeter wave cellular services for television and mobile communications, low earth orbit satellites, wireless PABXs, and PCS all suggest key new wireless markets. This is not to say these new technologies will displace or retard the continued rapid growth of fiber technology. It does suggest, however, that free-space communications will continue as a counterpart or complement to fiber telecommunications. Further this dynamic balance is likely to continue some years into the future. The case of TCI together with US WEST providing "combined" cable television and telephone services in the United Kingdom is only one case in point. Equally significant is AT&T's pairing with McCaw as is the case of Motorola buying a large share of NEXTEL. All of these developments and more suggest that "hybrid systems" will be increasingly likely around the world. This same integrated "wire and wireless"-based solution can also be expected as a fallout of the various mergers of U.S.-based regional Bell operating companies, interexchange carriers, equipment and software suppliers, the cable television organizations, and the rest of the so-called C-5 industries.[16]

3.3.6 Patterns of Replacement of Buildings and Infrastructure

The typical depreciation life of houses and buildings is 20 to 30 years. The practical life of automobiles, appliances, and even formal education is today about five to ten years. The "life" of most contemporary computers, information technology, and telecommunications is shrinking and is often only two to three years. In general, we are becoming more and more of an information society and more and more dependent on "smart" equipment and intellectual prosthetics. The problem of technological obsolescence for both equipment and software is not only increasing, but seems to be progressing at an exponential pace.

The problem becomes particularly acute where "systems" such as houses and buildings, which have traditionally had long lifetimes and leisurely depreciating schedules, suddenly intersect with "intelligent systems" that seem to have lifetimes measured in the blink of an eye. For nearly a decade, the concept of "intelligent buildings" has promised the prospect of broad band information services to the desk-top as well as such "smart" features such as 24-hour-a-day energy control systems, automated security systems, and so on. These systems have, however, turned out to be rather expensive, geared to standards that in some cases have become obsolete and are rather static in architecture.[17]

The new "smart" office buildings and soon "smart" home offices may be increasingly oriented toward wireless technology. This is based on a combination of cost effectiveness, cordless access to all telecommunications and information devices, reconfigurability, and the ability to provide band-width on demand. The only major issue seems to be whether microwave or millimeter wave, or perhaps even infrared will be used for this purpose. Since the atmosphere is the transmission media for wireless, it at least seems unlikely that it will quickly become obsolete.

In light of health considerations and potential broad band service requirements it is possible that these wireless service requirements will indeed migrate toward infrared services. A typical configuration in a high-rise office complex may very well become a hybrid of broad band monomode fiber in the risers and infrared or radio buses serving each office bay area.[18]

3.3.7 Public Policy Guidelines for Redundancy, System Availability, and Security in Communications Systems

It is often believed that a communications system is almost totally defined by basic concepts of band-width or digital throughput, telecommunications application, quality of service, protocol, frequency, transmission links, modulation and encoding systems, network size and node characteristics, and signaling and switching modes. These factors are indeed critically important, but there are several other factors that must be considered.

These are the factors associated with system availability, system redundancy, and security in communications systems. In a world wherein the field of telecommunications is developing increasingly on the basis of commercial competition and a world where military strife and rivalry is lower than in many decades, there is some tendency to overlook or minimize these concerns and issues. The incentive to have higher profits and to achieve lower operating costs in a competitive market can lessen or diminish concerns to achieve "extra margin" and higher levels of system availability and system security.

In the world of "wire" telecommunications one can now find regional or local pooling by interexchange carriers on one trunk. In such cases private networks can even be switched to back up the public switched networks in certain operating areas. In many other cases the private networks, however, are dependent on the PSTN, which are providing the "pipe" as a virtual system. The equivalent practice of not having back-up capacity can be found in the world of wireless when there is sharing of common towers to provide cellular service or even actual sharing of a rural and remote cells among competitive systems. In the coming world of personal communications service it is entirely likely that node interconnection systems, or add-on repeaters to address multipath problems will also be shared facilities. This is not to suggest that reductions in operating costs and sharing are not desirable where feasible. Rather, it is to suggest that ISDN standards for system availability of 99.98 percent may be increasingly difficult to achieve if levels of redundancy and back-up facilities are not provided. Furthermore, current regulatory trends could actually result in large-scale problems when single point of failure problems occur and little attention has been given to the need for redundancy and backup systems.[19]

There are several key factors to bear in mind with regard to such issues. These are as follows:

(a) Wireless communication, because it relies on free space rather than a physical conduit has an increased opportunity to survive a disaster. This is especially true of floods, fires, earthquakes, volcanoes, tornadoes, and hurricanes.[20]

(b) Security is a much more complicated matter. Digitally encrypted systems are probably well secured regardless of the transmission media. (Some feel that fiber is much more secure than free-space radio transmission, but the fact is fiber can also be intercepted—in at least six different ways. Radio obviously can be most easily intercepted but then there is no false sense of security engendered, unlike in the case of fiber.) Meanwhile, public concerns about protection of their privacy is at an all time high. Surveys conducted as of mid-year 1993 show that 83 percent of those surveyed were concerned about their personal privacy and 53 percent were very concerned. Comparable figures for 1978, some 15 years earlier, showed those with concerns at 64 percent and those who were very concerned at 31 percent.[21]

(c) The tandem use of fiber and wireless technology is among the best ways now available to ensure redundancy and back up of telecommunications facilities, and as such "wire" and "wireless" redundancy or "hybridization" could be helpful in ensuring high system availability levels.[22]

(d) National policy and regulatory guidelines to ensure effective backup and restorability of telecommunications services for both public and military purposes have been developed by the U.S. Department of Defense, the National Telecommunications and Information Administration, and a high-level U.S. industrial panel. These, however, need to be strengthened and kept constantly up to date. In the age of the information highway this will become even more critical.[23]

3.4 WORLDWIDE MARKET ASSESSMENT OF WIRELESS TELECOMMUNICATIONS

As reflected in Figures 3.3. and 3.4, it can be seen that the wireless telecommunications market is really in no way monolithic and, in fact, it is divided into many discrete submarkets. Despite this segmentation, there are certain key macroissues that characterize the global trends with regard to wireless communications.

The world of telecommunications is becoming increasingly more international and interconnected. In spite of this very definite globalization process, however, there is no uniformity of practices in this field around the world. Today there are major inconsistencies in the allocation and use of frequencies for wireless communications and significant differences with regard to the standards that are in use. Frequencies used for television and mobile communications are still used for long-distance HF and VHF communications in the so-called developing world. There are many different formats or standards for mobile communications that will be discussed further in later chapters. There are important standards differences with regard to modulation, encoding and multiplexing techniques, cell size, frequency reuse schemes, and channelization.[24]

On a worldwide basis, Global System for Mobile (GSM) has won broad acceptance as a digital cellular service. This GSM standard uses Time Division Multiple Access to provide service throughout Europe and many parts of Asia. It is now established as an efficient and reliable service that can serve both vehicular and pedestrian modes of traffic. In North America, however, there are a number of options that include Specialized Mobile Radio, analog mobile phone service (AMPS), and digitally based systems that especially include the IS-54 standard in the United States. Most recently the Joint Technical Committee adopted multiple PCS standards that include both TDMA and CDMA systems. As has occurred many times in the past, there is a clear prospect that the world wireless market will be divided by different standards. These standards primarily differ on the basis of frequency allocations, modulation, encoding and multiplexing methods, and operating systems. In the past, such divisions might have implied that local or regional manufacturers of equipment had promoted the difference in standards in the hope of becoming or remaining market dominant in their area.[25]

This is no longer necessarily the case. World-class manufacturers with global trading patterns are adept at manufacturing, marketing, distributing, and servicing equipment and products engineered to local or regional standards. Many now manufacture devices that at the click of a switch can convert from one regional standard to another. In short, telecommunications standards, although they vary in different regions of the world, are today much less of a trade obstacle. For decades the nontariff barrier was almost as large a barrier as tariffs and duties, but because of the increased technical agility and flexibility of manufacturers and changing regulatory contexts, this is now much less true.

Further, with the new GATT agreements that enter into effect in 1995 along with the creation of the new World Trade Organization, this trend toward reduced restrictions on global trade is likely to continue and perhaps even improve. One of the areas that has been almost totally opened to competitive international trade is that of information technology and telecommunications. Although some restrictions on defense-related technologies remain in effect, these too are being reduced. The total global level of trade in telecommunications services is as of the end of 1994 approximately $550 billion and is expected to grow to $1 trillion by the year 2001. Of these amounts almost one-third represents global trade. Further, the trade balances between and among Europe, Japan, and North America in the telecommunications equipment and service sectors are close to being equal—certainly much more so than in most other critical trade categories. Thus the telecommunications field represents a very large market well suited for active competition and not beset by protectionist concerns.

Nevertheless, this field does represent some complex trade issues. In particular, Japanese restrictions in allowing free and open access to their markets for cellular telecommunications devices have been the explicit cause of a trading crisis. Formal complaints filed by the Motorola Corporation to this effect stimulated one of the most protracted disputes of the 1990s between the United States and Japan.

3.5 KEY MARKET TRENDS

In coming years wireless technology and applications will likely continue to develop vigorously. In this process there will likely be a pattern of overlap, merger, and hybridization, so that the strengths and weaknesses of wire and wireless transmission systems can and indeed probably will complement each other. In projecting future market trends it is important to start with some well-documented guidelines. First of all, there is a historically documented pattern of telecommunications development wherein revenues for telecommunications services in general, and for wireless technology, in particular, tend to range from 40 to 65 percent invested capital . Most typically, revenues for wireless services can be expected to remain closer to the 65 percent end of range for gross invested capital. Figure 3.7 shows a projection of the growth of both wireless telecommunications revenues. This pattern generally seems to hold true even in what might be called periods of technological discontinuities or major transitions.

Even the start of cellular radio telephone and the beginning of personal communications service do not greatly distort the 40 percent rule, probably for two reasons. New technologies and applications at the outset represent a small percentage of the total revenues or investment. Further, there is significant "smoothing" of investment by using longer-term financing arrangements to eliminate heavy spikes of investment. Other key rules of thumb for the telecommunications industry, within the United States at least, is that sales per employee often range between $150,000 and $230,000 per year, while profits per employee per year would often fall between $10,000 and $18,000.[26]

Pay as you go financing is thus often used in well-established telecommunications services because of the strong revenue stream. In new services or technologies, however, even if there is a large revenue stream there is still typically a need for much more intensive capital financing. This relationship between capital financing, service growth, and rates of innovation are crucial to understanding today's telecommunications world. In the past, telecommunications operated under an environment of rate base regulation. Investments accepted into the rate base were then used to establish how much profit could be earned. These capital investments were often financed over 15 to 20 years, and the process, in effect, ensured the long-term profitability of the telecommunications operators. But in today's

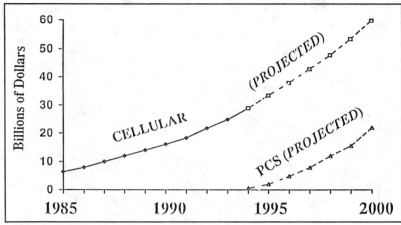

FIGURE 3.7 Projected Cellular and PCS Revenues

fast-paced world of digital technology, with rapid innovation and short half-lives for equipment and software, the period of amortization is much more reasonably three to seven years in length. New enterprises coming out of a background of the computer industry are often much better able to adapt to this environment than those that have evolved from the monopoly telecommunications background.[27]

Another guideline that can be used to project future markets for wireless telecommunications is simply the application of the "S" curve rule. This profile of take off, with rapid exponential growth, tapering off after a few years and moderating to slower growth is the classic "S" curve. It characterizes almost all markets for successful new products and services and also most biological systems. In some cases, however, there is a second or even third phase of development with major innovations in the product or service allowing a series of "S" curve waves to occur. The evolutionary stages of television from black and white sets, then color television receivers, then wide-screen projection units, and perhaps ultimately, high-definition television, is perhaps a classic case in point. The patterns seen in Figure 3.8, which could be characterized as a parallel trend analysis might be considered analogous to the development of the initial analog cellular radio telephone service followed by digital GSM cellular and then PCS services.[28]

Another key market trend is that related to the latest management techniques and modern management information systems. It is sometimes assumed that market success and especially market dominance is based on superior technology. In today's global markets with large corporations funding sophisticated R&D programs, complex standards-making operations, and a myriad of start-up venture capital firms often developing new leading edge technology, it is increasingly difficult to dominate telecommunications markets. This can today seldom be done through technology alone. The key is, more often than not, innovative management and marketing rather than simply having the best technology . The most successful telecommunications corporations are usually those that are best at identifying the market, the customer, and especially the customer's wants and needs.

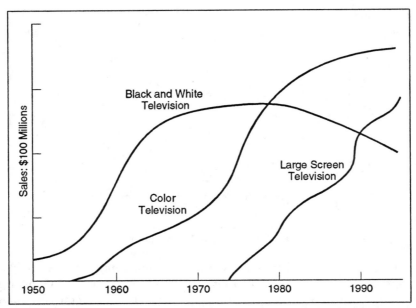

FIGURE 3.8   Parallel Trend Analysis: Historical Patterns of Television Sales

3.6 THE WIRELESS MARKET: A SECTOR-BY-SECTOR ANALYSIS

In light of the diversity and complexity of wireless services it is essential to break this market analysis down into its key constituent parts in order to review major trends for each service area. In particular, it is useful to review the following sectors: (a) paging; (b) cellular; (c) specialized mobile radio; (d) dispatching, public safety, and emergency mobile communications; (e) position location and determination services; (f) microwave relay; (g) over-the-air radio and television broadcasting; (h) satellites; (i) personal communications services; (j) wireless LANs, PABXs, and intrabuilding buses; and (k) future services (i.e., SHF cellular television distribution or LMDS, infrared buses, etc.).

Paging.   There is a current tendency to discount the future growth and development of this service. The oft-expressed question is: In a world where everyone has personal telephone service, why would anyone want or need a pager? Yet in a world of personal computers there are still hand-held calculators, electronic calendars, and digital schedulers. Matches were not made obsolete by cigarette lighters and power mowers have not eliminated the manual variety.

The new market areas into which conventional analog pagers may diversify as personal communications becomes more prevalent may well include: (i) paging services in developing and industrializing countries where mobile personal communications is very limited in its availability or is very costly or both; (ii) users of paging services who wish to limit their availability and control how their time is used; (iii) users who prefer pagers to personal communications for reasons such as reduced cost, personal style, and so on, and (iv) paging as an emergency backup.

The advent of new more capable two-way messaging services via narrow band Personal Communications Service (PCS) will also tend to redefine the mobile messaging services. The new narrow band PCS service at 900–901 MHz, 930–931 MHz, and 940–941 MHz will be able to send text return messages very easily and cost effectively. The fact that national bidders in the U.S. market were willing to bid hundreds of millions of dollars on these frequencies suggests that there many well be billions of dollars of future revenues in this new interactive paging service.

In the very short term, it seems likely that in terms of dollar sales volume the pager market will be substantially overtaken by wireless communications. The actual number of pagers may, however, exceed the number of personal transceivers even though the number of users of one-way and two-way service will likely grow closer and closer together. The great assets of pagers will remain low cost, compactness, and user control. Features that are likely to grow in importance over the decade are "silent or vibrating ringing," enhanced digital readout messaging, fax- and e-mail-related capabilities, and automatic number identification supported operations. The greatest single advantage of pagers is the simple fact that not everyone wants instantaneous two-communications at anytime, anywhere, or with anybody—especially anybody.[29]

Cellular Telephone / Fax / Data.   There are some "technophiles" who believe that not only pagers, but also analog cellular telephones will quickly go the way of the dinosaur. Their logic is quite straightforward. If one can have what many believe to be high-quality digital service, why would anyone wish to remain with inferior-quality analog cellular? Again, in real life the equation is never that simple.

There are, in fact, a host of factors to consider. Sometimes, one of the most mundane issues can still be crucial. In the United States, the simple truth is that the supply of new telephone numbers is dwindling. Today the local exchange carriers and the cellular telephone operators have numbers, but the new personal communications service operators do not. It is likely that by 1995, when PCS comes on-line, telephone numbers will not be available at least in some markets. Furthermore the new Service Access Code Numbers (The 500 series) could also be in short supply. It is hard to sell PCS services especially when you have to buy that frequency at high cost. But when there are no telephone numbers to assign to the user it is very difficult indeed. It will take time for the FCC to not only complete the auction of the PCS frequencies but supervise the transition of existing users to new frequencies and perhaps even more time to get a major supply of new telephone numbers. In short, cellular telephone operators have an enormous advantage in terms of a head start in marketing to carefully targeted business customers and consumers. Further, there are "squatters rights" on assignable telephone numbers, lower-cost frequencies upon which to operate, and presumably, lower cost service charges at least at the outset.[30]

The cellular industry clearly has the advantage that derives from being entrenched in the market in terms of customer base as well as installed equipment. This means in the U.S. market about $10 billion of plant investment, nearly 15 million subscribers, and good progress toward having Number Seven Signaling connectivity for roaming services. There is also the advantage of "name" recognition and large capital resources. In the United

States, for instance, Mobillink and Cellular One both have a great deal of name recognition and a good deal of economic and political clout. This is particularly true now that AT&T has purchased the McCaw Cellular system. New entries into PCS will need to advertise extensively to overcome this disadvantage. To the extent that existing cellular operators extend their operation into PCS, many of these difficulties can be overcome on an internalized strategic planning basis.

In addition to these market, regulatory, and economic reasons, there is also a key technology issue as well. In the United States, cellular service operates in the 800 MHz frequencies and PCS will operate in the 2 GHz frequencies. Just because of the different physical characteristics of the radio waves at these two frequency bands there are implications for the design and operation of the two systems. PCS for the most part comes out the worst in such comparisons. The problems of multipath are more difficult for the smaller wavelengths used for PCS. The mandated power levels are also less for the higher frequencies. The bottom line is that PCS cells will need to be smaller and thus more plant and equipment will need to be installed. Furthermore, the smaller the cell size and the lower the power, the more difficult it will be for PCS to offer mobile services to automobiles and trucks.

A further technical factor that will have a key impact on the market is that of the size, cost, and performance of hand-held transceivers. Major progress continues to be made in this area. Recently it was announced that a five-chip transceiver could provide all mobile telecommunications and pager functions and that these components could all operate at 3 volts. This means that the hand-held unit can be made at very low cost and that the battery unit can be greatly reduced in size and performance. Furthermore, these units can operate at low power levels for voice service and for pager services at ultra-low wattage. These new units, which are available now, should further entrench the conventional cellular services over the next two years.

These differences between PCS and cellular, however, should not be overemphasized. The first concepts of PCS tended to equate this service with microcellular or even picocellular service involving only very tiny cells. More recently, however, market and technical studies for PCS are tending to recommend designs that have much larger cell sizes. The latest iterations of PCS designs are tending toward operating and performance characteristics that closely resemble conventional cellular. This shift to "look" more like conventional cellular is to counteract the problems noted above and to reorient the service in order to provide service to vehicles.

The crucial market factor between conventional analog cellular service and PCS may well reside in the area of digital cellular services. Currently conventional cellular systems are poorly suited to fax or digital services with frequent errors, disconnects, and retransmissions. The cellular industry is making major efforts to provide a high-quality analog-based data transmission system known as Cellular Digital Packet Data (CDPD), which will offer a robust data channel extracted out of available "set-up and downtime" channels. Estimates are that this service will be available during 1995 and overcome the most important market liability that conventional cellular has with regard to the new PCS service offerings now envisioned in 1996 and 1997.[31]

Specialized Mobile Radio.   The market represented by Specialized Mobile Radio is in many ways a remarkable and almost unexpected development. In the mid-1980s specialized mobile radio service was largely a hodgepodge of small businesses offering fleet mobile services using a crazy quilt of frequencies. It was not big business or exciting or particularly lucrative. With the enormous success of analog cellular radio services, however, some very enterprising entrepreneurs concluded that this might be an interesting way to enter the exciting new mobile services industry through the back door. These new-look SMR firms started to buy up and consolidate small firms with existing licenses in various key metropolitan regions. As meaningful capital was raised to finance several of these enterprises, the picture changed quickly. A significant number of licenses for mobile radio frequencies were purchased in this manner at a small fraction of the cost associated with acquisition of cellular systems. Subsequently, new and much more up-to-date radio mobile technology was applied to increase the effective capacity by a factor of ten or more times. This updated SMR service is frequently referred to as the "enhanced specialized mobile radio" service or ESMR.

The aggregated strength of the new ESMR companies, especially NEXTEL, is very strong. The ESMR companies have generally acquired their licenses at a cost per population some three to five times less than the cost of acquiring conventional cellular. This ultimately attracted favorable attention in the financial community. When some of these firms, which had spread out into regional and even national enterprises, incorporated and offered stock to the public, they became very hot properties. The first of these, FleetCall, which is now known as NEXTEL, actually became instantly valued at over $1 billion with its first public offering.

Today the ESMR industry is well established in the United States as an "alternate" form of cellular radio service and is growing at a rapid pace. New frequencies are thus being opened up for mobile services in SMR, ESMR, and PCS bands. This suggests that these new and emerging companies will quickly become "established" players in the personal communications services business. It is possible in the digital mobile communications services market that ESMR will be integrated into this broader category so thoroughly that this historically important service description and distinction may even fade away.[32]

Dispatching and Emergency Mobile Services.   If one were to describe the most important changes in how police, fire, ambulance, rescue, and even military operations have been carried out over the last three decades, communications would be a key part of the story. Communications and computer support systems have been employed to speed response times, redeploy resources on demand, and to carry out remote or decentralized operations. Operations and decisions, which would have once been delayed until an expert could address the issue at a centralized site, are now decided in the field. This pattern has been repeated many times over. Paramedics with telecommunications and computer links to hospitals carry out emergency treatment procedures. Police patrols can trace stolen cars, identify suspects, or reroute traffic with mobile computer network links to local precinct offices or even links to FBI or INTERPOL mainframes. Army units can access GIS data systems to learn about local terrain or use hand-held GPS receivers to know their location to within a few feet. They can also communicate directly to air reconnaissance pilots.

The key in all of these innovations has been to deploy "smarter" and more interconnected remote units and to extend field operations by giving instantaneous intelligence to headquarters. The enabling technology in this respect has not been higher-performance computers or upgrades of headquarter systems, although these innovations have occurred. The ability to provide wider band, more reliable, and more cost-effective mobile communications has truly been the key to this revolution. The results have been impressive. Better communications has led to faster response, better and more accurate performance, and more effective use of personnel. It has also allowed better use of expertise and the near-instantaneous ability to deploy emergency, disaster recovery , firefighting or military services. It has in short allowed more to be done with less. The combination of new enhanced 911 services with upgraded mobile communications services have allowed lives to be saved, property loses to be reduced and more effective use of limited human resources to be achieved.

This trend can be expected to continue as various emergency and military operations are designed to work "smarter" and "leaner." Rapid growth in these markets can be expected. In this case, the growth will not so much involve more mobile units, but rather the upgrade of mobile units to be much more capable and to be much broader band in the types of image, data- and video-based services that are offered. High-definition medical information, fingerprint images, case files, traffic routing schematics, and navigational and mapping tele-services to vehicles will become routine. Other key elements in these new and expanded remote mobile services will also involve complex digital encoding, antijamming systems, and counter and even counter–counter measure communications devices to deal with problems of security, reliability, and even disruptive and invasive tactics designed to disable the critical communications systems. It is also likely that more conventional dispatching radio services will be used in parallel with cellular telephone services. A number of police forces have already experimented with a combination of dispatching radio and cellular telephone-equipped patrol personnel in crowd control and riot situations and found that such hybrid systems were more effective than relying on just one system, especially when dealing with a crisis in "congested" time periods.[33]

It seems entirely possible that the total volume of communications in these areas, both in the United States and abroad, may expand in terms of total digital throughput by a factor of ten by the year 2000.[34]

Position Location and Navigation. The single biggest improvement in the performance of any communications service, measured in relative terms over the past decade, most likely would be in the area of position location and navigation systems. Today position determination can be accomplished at a cost that is an order-of-magnitude less than in the 1980s, while the result may be two to three orders of magnitude more accurate. One can today instantaneously find out one's location, thanks to the 28-spacecraft Global Positioning Satellite (GPS) system. Armed only with a $700 GPS hand-held receiver one cannot only learn one's own location but also the rate of closure to another reference location. If the automobile industry had made similar breakthroughs in the last decade one could buy a limousine for about $1000 and obtain about 1500 miles to the gallon in gasoline mileage. Part of this comparison is not entirely fair in that the space segment is provided by the U.S. military without any charge to users. If there were somehow to be a user fee imposed (recognizing that this would be very difficult since this is a receive-only service), the costs would clearly be much higher.

This navigational and position location capability represents another high-growth wireless telecommunication service that can be offered to both fixed and mobile users. Within the next five years, many believe that over half of all new cars sold will be equipped with GPS receivers for navigational services. Also, personal navigational devices for a limited population of emergency and military users will rapidly expand as well to become a general consumer item as the cost of GPS receivers drop below $500 and become wristwatch-sized. Each year new applications are being found for GPS receivers. These include use by hikers, cross-country skiers, commuters, tourist navigational systems, rental car services, truck fleet management, firefighters, peace-keeping personnel, and military operations. One of the most innovative new GPS services include low-cost ways to obtain environmental data about the earth's atmosphere. This service is expected to be provided by ORBCOM in 1995–1996.

What makes this service unique from a commercial market standpoint is that the military provides the satellite system that allows the precise tracking. Thus the key satellite service is provided "free" and completely outside of the consumer payment or commercial provisioning process. The market is thus entirely in the sale and leasing of GPS receivers and in software to make the GPS system easier and more precise to use. This alone could still grow to be a billion dollar business in the next century.[34]

Microwave Relay.   The previous brief market assessments might have given the impression that every conceivable wireless telecommunications service is in high demand and that rapid market growth can be expected in each and every area. Microwave relay, however, represents a departure from this trend. Clearly, broad band microwave telecommunications relay is experiencing a decline that is likely to continue. This technology, even with the introduction of digital transmission techniques, is likely to become even more of a niche market. This niche is defined by rural and remote areas where the volume of traffic or the terrain and natural barriers rule out the installation of fiber optic cable. There will thus be a clear-cut but small market that will not even grow with the rate of inflation over the next decade.[35]

Over-the-Air Radio and Television Broadcasting.   As noted at the beginning of this book, this is a market segment of the wireless industry that is in trouble. For years, the overwhelming emphasis in the traditional radio and television industries has been on devising programming for its audience. The distribution systems via over-the-air broadcasting stations and national microwave distribution have been largely taken for granted. In the United States and abroad the emphasis has been on keeping the broadcasting systems running, licenses for frequencies renewed, and coordinating news feeds from local and regional sources readily available. The advent of satellite communications, fiber optic cable, digital compression technology, and cable television networks was often seen by television executives as more of a new wrinkle or a discomforting blip on the screen rather than a new opportunity. The extent of the liabilities associated with conventional broadcast distribution systems are only today being noted and understood.

The following comparative evaluation of the latest cable television system versus conventional television broadcasting is given below in Figure 3.9. It should be noted that many of these comparisons also apply to conventional radio versus cable and DBS-based radio.[36]

Service Features	Conventional Television	Cable TV
Channel Capacity	Limited	36–500 Channels
Quality of Signal/Audio	Variable	Very High
Unedited Movies	No	Yes
Movies on Demand/PPV	No	Increasingly Yes
Adult Programming	No	Yes
Nonstop News/Weather	No	Yes
Narrowcast/Educational Programming	No	Yes
Interactive Programming	No	Increasingly Yes
Value-Added Services	No	Increasingly Yes
Data/Fax/Telephone Option	No	Shortly

FIGURE 3.9   Cable Television versus Conventional Television

The comparison between cable television and conventional television as given in Figure 3.9 could be developed in greater detail and with greater precision, but the basic conclusion would be the same. Conventional television is at a disadvantage in terms of channel capacity, quality of video and audio signal, and in the range and type of programming it can provide. Conventional broadcasters also largely lack the ability to provide interactive services, and have a limited range of new service offerings (i.e., HDTV, security services, etc.).

Furthermore, there are limited new competitive options available to over-the-air television and radio networks and their local affiliates. The conventional broadcasters have huge sunk-investments in certain set frequencies, transmitters, power supplies, and so on, and these cannot be easily sold or upgraded to duplicate the capabilities of cable television systems. Available options are thus to invest in direct broadcast satellite systems, develop interactive terrestrial cellular systems using the millimeter wave frequency bands, or try to develop some form of partnership with cable televisions or regional Bell operating companies. There is, of course, the more limited option of focusing exclusively on the development and sale of programming and thus gradually abandoning the distribution and transmission role entirely. It is significant to note, in considering strategic options, that the Bell operating companies and the interexchange carriers in assessing their own commercial interests have shown far more interest in partnering with what are at least perceived to be the "high-tech" broad band cable television multiservice organizations rather than the "low-tech" broadcasters.

Many analysts indeed believe that broadcasters will opt to concentrate more and more on programming and perhaps ease out of program distribution. This would suggest that broadcasting facilities might start to fade away in favor of cable and DBS systems, but this seems unlikely to happen overnight. The value of the broadcast transmitters may diminish but important uses will likely remain. Although the so-called "big three" networks share of the national U.S. market has shrunk from nearly 100 percent in the 1960s to about 60 percent in 1994, the rate of decline is now much less rapid. Again, this is somewhat misleading in that the home movie-rental market, which has grown to over a $15 billion business, is not reflected in these figures. What is clear is that the television options are limited and that interactive cable television will serve to worsen the position of conventional broadcasters. The possible options for the over-the-air transmitters might include:

(a) Reconversion or cannibalization of existing facilities to support the provision of new terrestrial cellular services in the millimeter wave range as is now being demonstrated under experimental licenses.

(b) Conversion of existing facilities to educational or other narrow-casting uses with highly targeted audiences. This would allow concentration on mass-market entertainment programming that others would distribute.

(c) Development of a new hybrid or combination package that provides national multi-channel programming through DBS systems, but continues to use local affiliates to provide localism in major urban service areas. This might provide local news, weather, community sports, and so on.

(d) Conversion to digital transmission and compression techniques to offer multiple channels and pay-per-view and video-on-demand services. (This would probably involve rechannelization of existing UHF and VHF frequencies as well as the possibility of moving to the higher millimeter wave frequencies as in (a) above).

Clearly, some new strategies and partnerships are urgently needed if conventional broadcast television is to survive. Furthermore, the radio broadcast industry may soon also find itself under the threat of direct broadcast radio services. The conventional radio broadcasters will likewise need to adapt even though the competitive pressure will be later and much less than for the television broadcasters. Furthermore, as the newspaper industry and other information providers move to create new concepts, such as the electronic newspaper, wherein wireless updates of news are fed to electronic notebooks or tablets, the sense of competition and potential technological obsolescence in the broadcasting world will likely further increase.[37]

In the overall global market forecasts provided in Figure 1.8 it will be noted that the broadcast industry, unlike virtually every other sector of the wireless market, is not really expected to grow and may toward the end of the decade even start to decline. In terms of an assumed 5 percent per year inflation rate, a net revenue reduction is projected by the year 2000.

Today advertisers who support the broadcasting industry are diversifying their ad placement and this will likely continue. In the United States and abroad, more programming and programmers are seeking commercial support through new networks, new and existing cable television channels, and other new avenues such as DBS, lower power television, and so on. There will need to be rather profound changes if the broadcasting industry is to thrive or even survive into the next century. Strong political support, strong grips on existing licenses, the geographic extensiveness and local ties of affiliates, programming muscle, opportunities for new partnerships, and potentially, some technological innovation are still strengths of the broadcasting industry with which to be reckoned. These are tools and resources, however, that have limits. Some fundamental innovations will be needed within the next decade to ensure longer-term survival. Finally, the role of public radio and public television programming and distribution will likely need to be redefined and to an extent reinvented by the year 2000.[38]

Satellites.   The future of satellite communications is currently at a significant crossroads. To date the highly dominant source of all satellite revenues has been fixed satellite services for long distance, particularly international telecommunications services. It also includes a substantial amount of revenues from video distribution and for services to rural and remote areas. Business services to customer premise antennas, especially in the context of very-small-aperture antennas and the newly deployed ultrasmall-aperture terminals have also provided a significant source of growth over the last five years as the number of VSAT sites has skyrocketed. Sales in the United States have averaged nearly 25,000 per year since 1990 and international sales of VSATs have increased since 1990 from about 4000 per year to nearly 8000. As of the end of 1994, there were an estimated 160,000 VSAT terminals in operation on a global scale. These interactive VSATs were distributed 72 percent in North America, 10 percent in Europe, 8 percent in South America, 7 percent in the Asia-Pacific, and 2 percent in Africa and the Middle East.

In all these areas, fixed satellite services and related procurement of equipment represented perhaps a $12 billion a year industry. Furthermore, military communications satellite services, which is largely a "hidden" industry, and maritime communications services add another significant source of revenues and wireless telecommunications services. These may total another $7 billion in revenues and expenditures, albeit, not all of these figures in the military sector are clearly verifiable. In the coming decade, modest growth will continue in these established areas, but the areas of the greatest growth in terms of percentages will be in the new areas of land mobile, direct broadcast, and navigational services by satellite. In particular, mobile satellite services are expected to grow from $1.5 billion to $8 to $10 billion by the year 2001. Broadcast satellite services are also expected to rise dramatically from $1 billion to $10 to $12 billion by 2001.

Based on current plans in the United States, Europe, and Japan, billions of dollars in new capital will also be invested in new satellite systems for direct broadcast satellites such as the Hughes DirecTV, the N-Star of Japan, Astra -2, the next generation of BSB satellites and possible follow-on systems to the TDF system of France, the TV-Sat system of Germany, and the INSAT system of India. Further, DBS systems could even be initiated for several large developing countries. There could even be a global-based DBS system, and one or more additional systems in the United States, such as the Cellsat or Norris satellite system. To an extent INTELSAT has already begun leasing high-powered transponders for regional DBS services on its new high-powered INTELSAT VII satellite, beginning with a regional Arab television distribution system in approximately March 1995. Finally, several DBS radio systems designed to operate in totally different and much lower bands than those reserved for television are envisioned with at least six systems proposed to the FCC for the United States and at least two international DBSR systems are planned. In all, a fivefold to even tenfold growth in DBS markets seems likely in the next seven years.[39]

Equally dramatic is the potential increase in satellite systems optimized for land mobile services. Some of these would supplement existing terrestrial-based cellular systems while others would be aimed at the huge but geographically diverse rural and remote market for telecommunications services, largely in developing and industrializing countries. These systems actually come in a number of different design types. There are geosynchronous systems such as the American Mobile Satellite Corporation (U.S.) and the Telesat Mobile Incorporated (TMI) of Canada. There are medium earth orbit satellite systems such as the TRW Odyssey system and the INMARSAT Project 21 (now on hold at the instruction of the INMARSAT Council), which will likely be a combination of geosynchronous and medium earth orbit satellites. Then there are a variety of "big LEO" satellites in global constellations to provide voice and data services and the "little LEO" systems, which require much fewer satellites and investment costs, but can only provide store- and forward-data services. Although most of the systems are U.S.-based, there are proposed systems at various stages of planning from France, Brazil, and at least two from Russia. These systems are expected to grow into a multibillion dollar industry within the next decade.

What is key to understand is that the anticipated market for mobile satellites services has several different segments—each of which will have separate technical characteristics and rather different tariffing concepts. All of these satellite systems, however, share the need to make arrangements with national telecommunications agencies with regard to splitting revenues and obtaining authorization or "landing licenses" to operate in the affected countries. Some of the satellite systems also have an additional challenge of receiving permission to operate in frequency bands that are currently restricted for such use at the national or the international level. The various land mobile markets can be generally grouped as follows:

(a) Executive Mobility Service: This is a very-high-value voice and data service for executives who must travel to remote areas with very limited or no modern telecommunications. Estimated tariff: $3–4 per minute. Example: Motorola Iridium.

(b) Remote Mobile Service: This will typically involve extending cellular telephone service to low-density areas where terrestrial systems would not be economic. Estimated tariff: $1–1.50 per minute. Example: TRW Odyssey or Ellipso.

(c) Rural Telephone Service: This would seek to provide telephone and data service to truly remote and unserved areas, especially in developing countries. Estimated tariff: $0.25–$0.35 per minute. Example: Teledesic Mega LEO Satellite System.

(d) Store & Forward Data: This would provide data, e-mail, and fax services to isolated area within hours of the message being sent. Estimated tariff: $.01–$.001 per byte.

These combined market segments are expected to represent a total market at the level of billions of dollars per year in the early twenty-first century. This market will in the next few years be dominated by INMARSAT, AMSC, and TMI, all of which are geosynchronous systems. This will change around the year 2000 as a number of new "big" and "little LEOS" plus several medium and elliptical orbit systems enter service around that time. If market studies presented to the FCC by applicants and independent studies undertaken by the Gartner group, Booz, Allen and Hamilton, and Kalba International prove accurate the products and services associated with mobile satellite services could expand from $1.5 billion in 1995 to around $8–$10 billion by the end of the decade. This fivefold to sixfold growth in six to seven years could be among the highest in the entire wireless industry. Certainly the ability of the new low and medium earth orbit satellite systems being able to provide high-quality voice and data service with low latency are key service features. They could conceivably compete with fiber optic performance if they can keep their costs down and get their throughput performance up to a higher level. Certainly, in mobile services, satellites possess a strategic edge.[40]

Personal Communication Services.   This is the area where a tremendous spurt of activity in the wireless field is now widely anticipated. Some believe this will be almost a self-fulfilling prophecy. The enormous size of the investment required to buy new frequency bands for this service in the FCC-sanctioned auction process, plus the huge investment in building a prodigious number of microcells and then interconnecting each node with suitable switching equipment is not going to be easy. Many billions of dollars will be needed to propel this new service into being. Although there are quasi-PCS services that already have a head start, such as IS-55, GSM, DECT, and Telepoint 2+ services, PCS starts well back of the now firmly established AMPS cellular telephone system. The comparisons of the two services were, in fact, noted in the previous subsection. Based on various market studies conducted around the world, digital cellular is today less than 10 percent of the global cellular market (actually under 5 percent of the U.S. market) and by the year 2000 PCS may still be only about 20 percent of the cellular revenues and perhaps 25 percent of the total users.[41]

Action Information Service has projected that between 1995 and the year 2005, PCS subscribers will grow from nil to about 17 million. Given the many unknowns, however, PCS could easily grow at a slower rate. Most analysts do not project that PCS will actually catch up to the conventional cellular service until well into the twenty-first century. Critical in this regard will certainly be pricing. Today cellular services in the United States average around $70 per month, while PCS prices are projected be in the $45–$50 per month range. For the reasons noted above, it seems quite likely that cellular will be able to match or underprice PCS when this service is ultimately introduced in 1997. The projected growth of PCS subscribers is provided in Figure 3.10, but this is seen as an optimistic estimate.[42]

Wireless LANs and PABXs.   If there is a market sector in the overall wireless communications field where the future is most unclear this is most likely it. For at least two decades or more LANs and certainly PABX switches have been wire-based technologies. Wired connections have been so pervasive that the idea of wireless technology seems somehow strange or foreign. The conditions, however, have changed. They have changed in terms of new wireless technologies that are now available and market demand for highly flexible, reconfigurable, and mobile services have led to serious consideration of both wireless LAN and PABX capabilities in the office environment. Many planners of new office complexes are now thinking in terms of fiber or coaxial cable as the trunking system in office tower raisers, but are specifying radio systems as intraoffice and interoffice connections to the desk-top.[43]

Future Services.   The future rapid development of wireless services into the next century seems very likely in that both growing market demand and new technology seem

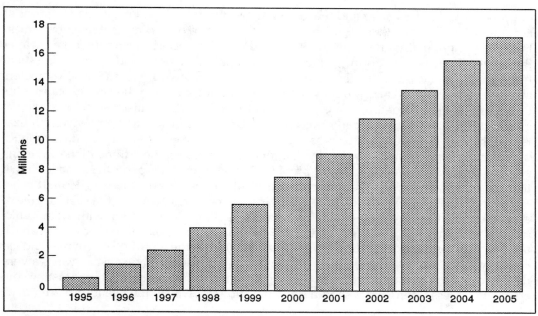

FIGURE 3.10   Projected Growth for PCS, 1995–2005.

to both be available in abundant supply. The possibilities for new services are severalfold. They include infrared buses that could provide highly flexible and reconfigurable broad band video, imaging, data, multimedia, and voice to the desk-top or to any mobile position within large office complexes in the context of both LAN or PABX services. It is certainly possible that infrared-based LANs and PABXs could reach a billion dollar market by the turn of the century.[44]

The services of the future also include picocellular digital links that would have sufficient capacity and intensity of frequency reuse to provide universal service on a global basis. This could in theory mean that every subscriber in the world could have mobile telecommunications service and thus also allow the implementation of the so-called universal telephone number to individuals as opposed to locations. It has been estimated by the International Telecommunication Union that some $300 billion in new telecommunications investment will be made by the year 2000 in just the top ten markets of the world and that mobile will claim an increasing percentage of that total investment. This suggests a huge mobile market not only for the United States, but for the whole world as well.[45]

The satellite systems of the future, with truly intensive frequency reuse and lower orbits to reduce latency in the transmission path, could assume a very broad role in terms of broad band ISDN services provided via Asynchronous Transfer Mode switching systems. This is to say that the widely perceived view that fiber optic cable provides a much more cost-effective and higher-quality service vis à vis satellite services need not necessarily be the case by the 2005 to 2010 time period.[46] On the other hand, totally new types of technology could impact on the satellite world as well, particularly in the form of the so-called High Altitude Long Endurance (HALE) platform. This is a concept of stabilizing by means of propeller turbines or solar and fuel cell systems, multiple communications platforms at altitudes of 18.5 kilometers (13 miles) for the purpose of providing mobile communications, television, distribution, and so on. The idea is that these systems could be cheaper than rocket-launched platforms, could be easily repaired or retrofitted by returning them to the earth's surface, and that they could effectively cover large areas of 180,000 kilometers at a time. Experiments carried out by the Raytheon Corporation in the United States, plus tests carried out in Toronto, Canada (i.e., the so-called SHARPS project), and tests in Japan all indicate that these HALE platforms are technically feasible using microwave stabilization in the long run. Prototype tests have shown over 90 percent efficiency in converting RF power into electric power for the rotary propeller motors on the "drone plane" antenna systems. Tests with jet and reciprocal propeller drives in the later part of 1994 have shown that the more conventional designs could probably be deployed in 1997 or 1998. It is believed that such systems can operate with virtually no latency and with a high degree of cost efficiency. These systems could, for instance, operate in conjunction with systems like the Teledesic mega-LEO satellite systems to provide gigantic telecommunications capacities for both developed and developing countries around the world.[47] There may indeed be many other technical possibilities for the future, but just the above list suggests a very dynamic range of new opportunities. Clearly the mobility, flexibility, and broadcasting coverage of wireless technologies should serve this industry well in the twenty-first century.

3.7 THE WIRELESS MARKET: TODAY AND TOMORROW

The broad market survey and forecast presented in Chapter 1 that suggested that all forms of wireless telecommunications are today in excess of $100 billion and will likely exceed $200 billion by the turn of the century is supported by a broad pattern of growth in many service areas. Digital cellular, PCS, wireless data, specialized mobile radio, wireless PABX and LANs, fixed and mobile satellite services, direct broadcast satellite, safety and emergency communications, and military communications will all likely see rapid growth rates of 12–20 percent for several years to come. Some, such as DBS and mobile satellite service, could grow even more rapidly. Certainly several factors could serve to retard this growth such as regulatory restrictions on new technologies, new licenses, and new frequency allocations. Any downturn in the global economy could slow growth as could any major difficulties or anomalies in the international standards-making process.

In general, the present prospects seem very favorable to growth. Key factors are as follows:

(a) Continued liberalization of telecommunications markets will occur. This will make these markets more competitive and open. (This will generally occur on a global scale.)

(b) Standards for GSM and PCS will expand to include TDMA- and CDMA-based systems and then evolve toward new standards for so-called Universal Personal Telecommunications (UPT).

(c) Digital compression techniques, more intensive frequency reuse systems, and more efficient modulation, encoding, multiplexing and error control techniques will enhance service quality and reduce the cost of mobile services. Additional frequency allocations at higher frequencies will also help fuel this trend.

(d) Businesses and individuals will insist on the flexibility and convenience of mobile systems in terms of hands-free communications, virtual offices, and instant communications from any location whether fixed or mobile.

(e) The cost of all forms of wireless and mobile communications will continue to decrease, particularly in relationship to the general cost of living.

Just as the overall field of telecommunications is moving toward convergence of many major-market segments, the wireless market is also moving toward its own form of consolidation of technologies and applications. This is reflected in the graphic shown in Figure 3.11, which shows four different forces of convergence for the wireless field. In short, as services become more digital, more integrated, and more multipurpose, the level of convergence will likely increase.[48]

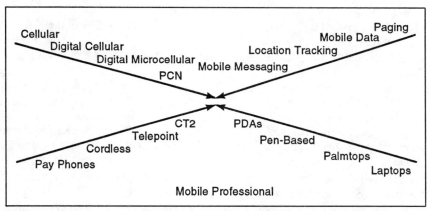

FIGURE 3.11   The Pattern of Technological and Service Convergence in the Field of Telecommunications. *Source*: Courtesy of the Yankee Group.

In general terms this pattern of rapid growth in wireless communications is about half dependent on economic, social, regulatory and policy issues on one hand, while the other half is dependent on technical and operational developments. Neither aspect can be easily ignored in any serious market assessment.

ENDNOTES

(1) Nakonecznyj, I.T., "The Wireless Revolution—It's Here Almost!" The NEC ComForum, Orlando, FL, 1992.

(2) "Global Trends in Telecommunications Services," Comquest, Nielson Information Services, September 1993.

(3) "The Communicopia Study: on Convergence" (New York: Goldman-Sachs, 1992; also see, "Future Telecommunications Markets," *US WEST Advanced Technologies Report*, April 1993.

(4) Center for Telecommunications Management, "The Telecom Outlook Report" (Chicago, IL: In-

ternational Engineering Consortium, 1994); also see Anderson, H., "The Mobile Professional" (New York: The Yankee Group, 1993).

(5) Kachmar, M., "The Goal Is Control of Time and Place," *Wireless*, November 1993, Vol. 2, No. 4, pp. 22-23.

(6) Toll, D., "PCS Takes the Field: Cellular's Concept of Personal Communications Services Wins Praise From its Users in Early Trials," *Wireless*, Vol. 2, No. 3, pp. 29-33.

(7) Taylor, J., "Trial by Auction: The Greening of PCS," *Wireless*, Vol. 2, No. 2, pp. 32-34.

(8) "Trends in Consumer and Advertising Spending in the Information Sector" (Denver, CO: Heintz-Barton, Inc., 1993).

(9) Mirobio, M.M., and B. Morgenstern, *The New Communications Technologies* (Boston, MA: Focal Press, 1990).

(10) Carey, J., and M. Moss, "The Diffusion of New Telecommunications Technologies," *Telecommunications Policy*, June 1985, pp. 145-158.

(11) "Global Satellite Television Trends," *Satellite Communications*, (Denver, CO: Argus Publications, April 1993), pp. 15-17.

(12) Carey J., and M. Moss, "The Diffusion of New Telecommunications Technologies," *Telecommunications Policy*, June 1985, pp. 151-152.

(13) Ingliss, A.F., *Behind the Tube: A History of Broadcasting Technology and Business* (Stoneham, MA: Butterworth, 1990).

(14) Carnevale, M.L., "Broadcasters Gain Support for Measure to Open Spectrum for New Services," *Wall Street Journal*, March 1, 1994, p. B6; also see Solomon, R.J., "Shifting the Locus of Control," *Annual Review of Communications and Society* (Queenstown, MD: Institute for Information Studies, 1989); also see Davis, J.A., "Cable Overbuild: Alternative Video Access Opportunity," *1993-94 Annual Review of Communications* (Chicago, IL: International Engineering Consortium, 1994), pp. 115-118.

(15) Vorick, F.L., "Cellular Service Evolution," *1993-94 Annual Review of Communications* (Chicago, IL: International Engineering Consortium, 1994), pp. 688-691.

(16) Cauley, L., "MCI's Entry Adds New Dimension to Wireless Race," *Wall Street Journal*, March 1, 1994, p. B4.

(17) Cross, T., *Intelligent Buildings* (Boulder, CO: Cross Communications, 1987).

(18) Pelton, J.N., *Global Talk* (Aaphen aan der Ryjn, Netherlands: Sijthoff and Noordhoff, 1981), pp. 84-87, 143-45, and 236.

(19) Wainwright, R.A., "Quality as a Competitive Edge," *1993-94 Annual Review of Communications* (Chicago, IL: International Engineering Consortium, 1994), pp. 939-943; also see Adkins, A.J., "Technology and Quality: Payoff to the End-User," *1993-94 Annual Review of Communications* (Chicago, IL: International Engineering Consortium, 1994), pp. 903-904.

(20) Dyer, J.E., "Disaster Recovery in a Cellular Service," *1993-94 Annual Review of Communications* (Chicago, IL: International Engineering Consortium, 1994), pp. 613-614.

(21) "Public's Privacy Concerns Still Rising," *Privacy and American Business* (Hackensack, NJ: Center for Social and Legal Research, September, 1993); also see Pelton, J.N., *Future View* (Boulder, CO: Baylin Publications, 1992), pp. 151-162.

(22) McCaw, C.O., "Cellular Communications," *1993-94 Annual Review of Communications* (Chicago, IL: International Engineering Consortium, 1994), pp. 43-44; also see ——, "Changing the World," *Wireless*, June 1993, Vol. 2, No. 2, pp. 23-26.

(23) "Global 2000 Report on Telecommunications" (Washington, DC: National Telecommunications

and Information Administration, 1990).

(24) Tsoi, K.-C.A., "User Interface Issues for Cellular Phones," *1993-94 Annual Review of Communications* (Chicago, IL: International Engineering Consortium, 1994), pp. 679-681; also see "Standards for Wireless Telecommunications," Tutorial, Supercom 1994, New Orleans, LA.

(25) Gustafsson, A., "Lessons Learned from Launching a GSM Digital Cellular System in Sweden," *1993-94 Annual Review of Communications* (Chicago, IL: International Engineering Consortium, 1994), pp. 618-621.

(26) Keen, P.G.W., and J.M. Cummins, *Networks in Action: Business Choices and Telecommunications Decisions* (Belmont, CA: Wadsworth, 1994), pp. 519-520.

(27) Brown, P., "Business Productivity Through Networking," *INTERNET*, April 1992; also see Weber, J., "Motorola's Rapidly Developing Peer-to-Peer Network is Helping to Achieve A World-Wide 'Wall-Less' Workplace," *Networking Management*, July 1992, pp. 14-16.

(28) "Strategic Planning and Market Analyses in Telecommunications" (Unpublished market studies for television, fax, EDI, and cellular telephone), University of Colorado at Boulder, 1993.

(29) "Motorola's Integrated Phone-Pager," *Wireless*, September 1993, Vol. 2, No. 3, p. 55.

(30) Lucas, J., "PCS vs. Cellular," *Telestrategies* (McLean, VA: Telestrategies, August 1993), pp. 1-12.

(31) Toll, D., "The Promise of CDPD," *Wireless*, March 1994, Vol. 3, No. 2. pp. 28-33; also see "CDPD—Here, There and Everywhere?" *Wireless*, June 1993, Vol. 2, No. 2, p. 6.

(32) Davis, H., "Enhanced SMR: The First PCS?" Telestrategies PCS Conference, Washington, DC, June 1994; also see "MCI to Acquire 17% of NEXTEL, an ESMR Cellular Firm," *Wall Street Journal*, March 1, 1994, p. A3; also see Berger, J., "SMR Shows its Utility," *Wireless*, September 1993, Vol. 2 , No. 3, pp. 43-36.

(33) Cole, L.S., "Phoenix Police Take Command of Wireless Technology," *Wireless*, September 1993, Vol. 2, No. 3, pp. 52-54.

(34) "AT&T Bell Labs Develops Intelligent Vehicle Highway System Toll-Collection System," *Global Positioning and Navigational News*, March 10, 1994, Vol. 4, No. 3, pp. 1-2; also see "The Japanese Automatic Vehicle Location Market Continues to Gain Momentum," March 10, 1994, Vol. 4, No. 3, Reference files.

(35) "The Changing World of Microwave Relay," *Communications Review: 1993* (New York: Upjohn, 1993).

(36) Ingliss, A.F., *Behind the Tube: A History of Broadcasting Technology and Business* (Stoneham, MA: Butterworth, 1990), pp. 1-60.

(37) Pelton, J.N., "Are Conventional Television Broadcasters Becoming Dinosaurs?" National Association of Broadcasters, Annual Technology Forum, Las Vegas, NV, 1993.

(38) Baer, W.S., "New Communications Technologies and Services," in Newberg, P.R., ed., *New Directions in Telecommunications Policy* (Durham, NC: Duke University Press, 1989), pp. 139-169.

(39) Marshall, P., "Global Television by Satellite," *The Journal of Space Communications and Broadcasting*, January 1989, Vol. 6, No. 4; also see "The DBS Market," *Via Satellite* (Potomac, MD: Phillips Publishing, 1993); also see "The Direct TV DBS System," *Via Satellite* (Potomac, MD: Phillips Publishing, 1993).

(40) Horwitz, C., "The Rise of Global VSAT Networks," *Satellite Communications* (Denver, CO: Argus Publishing, April 1994), pp. 31-34; also see Pelton, J.N., "Will the Small Satellite Market Be Large?" *Via Satellite*, April 1993; also see Special Edition on Mobile and Small Satellites: *Journal of Space Communications*, April 1993, Vol. 10, No. 2; also see Bull, S., "The Status of the Interactive VSAT Market," *Via Satellite*, December 1994, pp. 34-38.

(41) Toll, D., "PCS Takes the Field," *Wireless*, March 1994, Vol. 3, No. 2, pp. 18-19.

(42) "Studies Support Speedy Licensing," *Inside Wireless*, Vol. 2, Issue 7 (Englewood, CO: Four Pines Publishing, April 27, 1994), pp. 3-4.

(43) Kirvan, P., "Implementing a Wireless PBX," *Wireless*, June 1993, Vol. 2, No. 2, pp. 34-38; also see Palumbo, W.J., "Wireless PBX Access," *ICA Expo*, Dallas, TX, May 1994.

(44) Abe, G., "The Global Network," *Network Computing*, November 15, 1993, pp. 48-53.

(45) "World's Top 10 Markets for Telecommunications Investment Forecasted," *Global Telecom Report*, April 4, 1994, Vol. 4, No. 7, pp. 1-2.

(46) Manuta, L., "Big Leo Equals a Big Deal," *Satellite Communications*, April 1994, pp. 14-15.

(47) Glaser, P., "The Practical Uses of High Altitude Long Endurance Platforms" (Graz, Austria: International Astronautical Federation, October 1993).

(48) Kachmar, M., "The Goal Is Control of Time and Place," *Wireless*, November 1993, Vol. 2, No. 4, pp. 22-23.

Chapter Four

Wireless Communications Technology and Its Implementation

4.1 INTRODUCTION

The underlying basic technical concepts of the field of wireless communications were presented in introductory chapters one and two. Essentially, signals are modulated and sent through free space to receivers, where they are demodulated. These signals can carry any particular telecommunications service or even a combination of them such as voice, data, or video. Higher frequencies and broader band carriers are needed, however, for more complex and information intensive services such as video. In all cases there is a transmitter and a receiver, but the symmetry of this system can vary widely.

In cases such as a television or radio distribution system or a direct broadcast satellite system, the system is only one-way. In these cases there is often a single very powerful and high-elevation transmitter sending signals to a very large number of small low-powered receivers. In the case of cellular telephone or mobile satellite service there is a series of medium-powered transmitters distributed within multiple cells that are still asymmetrical but are of comparable power levels and usually interactive. Near the middle of each cell, a base station sends and receives signals within the cellular zone. These signals typically originate from lower-powered vehicular transceivers. Systems like wireless LANs and wireless PABXs are much like minicellular systems in that they also have transmitters at the cell center that connect to lower-powered remote terminals.

Finally, there are a number of wireless systems that are essentially completely symmetrical and nonhierarchical. These would include uniform networks where each transmitter and receiver is essentially the same. These would include microwave relay systems where a string of transmitters would be essentially the same size and power. The same type of symmetrical architecture would be found in a mesh-type interactive VSAT network. Other examples would be CB radios, walkie-talkies, and unlicensed PCS.

This chapter will concentrate on cellular mobile communications technology as broadly defined, while the following chapter will concentrate on satellites. This enlarged definition of cellular thus includes AMPS, IS-55, GSM, PCS, unlicensed PCS, and unlicensed Industrial and Medical Services (IMS), also known under FCC regulations as Part 15, SMR, and Enhanced Specialized Mobile Radio, and wireless LANs and PABXs. No attempt will be made, beyond the general background discussion provided in the introductory chapters one and two, to address amateur radio, microwave relay, CB radio, or television or radio broadcasting. This is because these sectors involve no unique technologies in terms of transmission, switching, signaling, or reception that will not be generically covered in the following pages. Also, the regulatory process for obtaining access to spectra to operate these services is well defined and has not been altered significantly in the last decade. Furthermore, the wireless mobile and the satellite markets are clearly the growth sectors of the next decade and are thus also commercially more interesting.

The fact that wireless systems are likely to become closely linked and integrated to terrestrial networks, especially fiber optic cable systems, has already been well noted. In some cases it is likely that cellular mobile systems will be integrated with terrestrial telephone or cable television networks for the purposes of cellular node interconnection. It is beyond the scope of this book to provide a detailed technical and architectural presentation of these wire and wireless integrated networks. It is nevertheless important to note that wireless networks of the future will not be islands onto themselves but interlinked in many complex ways to terrestrial wire systems. This will complicate greatly signaling and switching schemes. The asymmetrical nature of limited band-width signaling channels and broader band terrestrial signaling may, in fact, become a key problem in establishing twenty-first century hybrid nets within the PSTN.

4.2 WIRELESS MOBILE COMMUNICATIONS TECHNOLOGY

The two primary keys to modern wireless mobile communications technology are the creation and flexible interconnection of a large number of cells and the deployment of new digital technologies to create more channel capacity within the derived cells. These digital technologies include advanced multiplexing concepts (i.e., TDMA or CDMA), advanced digital encoding involving multiphase encoding of information, and digital compression techniques to allow high-quality voice at lower bit rates. These factors combine to enable wireless mobile communications systems to become ever more efficient and to handle more and more throughput within a fixed amount of frequency allocation. These technologies and how they are used to increase performance as well as to overcome problems of interference, multipath, and other related problems are addressed in the following sections.

4.2.1 Cellular and Unlicensed Wireless Systems

The idea of a cellular system of an unlicensed wireless communications such as an IMS service is quite simple. One creates a number of well-defined cell shapes—usually hexagonally formed—so that one can electronically isolate the frequencies used in one cell from adjacent cells. This concept is the same in an analog or digital cellular radio telephone system or in a wireless network within a corporate building. This staggering of the cells' frequencies allows reuse of the available frequencies within the many cells formed within the system. Those cells that use the same frequency, in order to avoid interference with nearby cells, must be physically removed from one another. The reuse patterns shown in Figure 4.1 are based upon an analog system. This follows a rule of significant separation by over two cell locations. With modern digital techniques as discussed later, much more intensive patterns of reuse can now be achieved, but still some separation is necessary. This basic concept of manyfold frequency reuse within the individual cells of an overall cellular system greatly expands the available channel capacity within a system.[1]

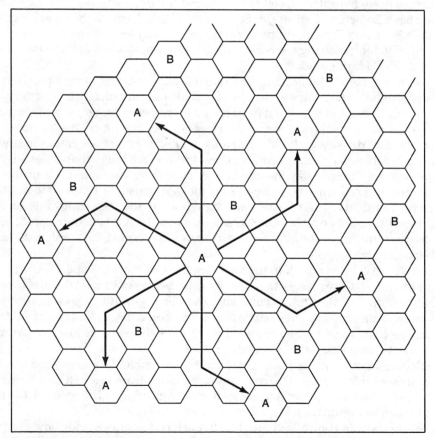

FIGURE 4.1A   Illustration of the Determination of Co-Channel Cells

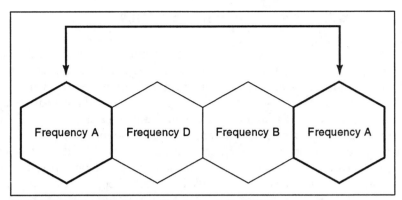

FIGURE 4.1B   Cell Reuse Scheme—3-Cell Diameters

There are several additional technical concepts that are important to introduce at this time. These are the definitions of the boundaries of cellular systems, hand off of mobile calls that cross these boundaries, cell splitting, and the detailed operation of Mobile Telecommunications Switching Offices (MTSO). The MTSO in terms of the switching hierarchy of the Public Switched Telecommunication Office is considered to be an extension of the Class 5 local switching exchange. It is because of its special status that it is considered a part of the overall PSTN framework.[2]

The ideal concept of a cellular boundary is usually represented as very precisely defined hexagonally shaped radio beams where the signal drops dramatically from a very high flux density to virtually nil and the signal from the adjacent beam center does the same. In fact, actual test measurements show that the ideal uniform beam with crisp edge does not really exist. Also, the flux density within the cell area may have creases or cracks in it where the performance levels are several decibels below the nominal and that other high-power-beam levels can be carried over the cell boundary into at least one-third of the adjacent cells. This imperfect cell boundary often limits the performance of the cell to only about 70 percent of the nominal capacity. There are many factors that can contribute to the "non-nominal" beam shaping and boundary creation. These include such disparate reasons as traffic patterns, terrain, buildings or other structures, trees, heat, rain, antenna design, and power shortages or outages.

Not all cells are created equal. In urban areas, base stations are operating at 500 feet or below and at e.i.r.p. power levels that are typically between 10 to 100 watts even though the FCC allows up to 500 watts. In rural and low-density traffic areas elevations of up to 1300 feet for mountaintop cell sites can be found and power levels that can be up to the 500-watt limit. The overlap of higher-powered rural cells and smaller lower-powered urban cells can also constitute major interference problems.[3]

This lack of clear and precise boundaries can create problems of degraded service, especially near the boundary lines. The problem of hand off from one cell to another can certainly be affected by the boundary performance conditions. The system to determine where and when to perform a hand-off is illustrated in Figure 4.2.[4]

There is one cell-site controller in each cell. Each cell operates under the control of the Mobile Telephone Switching Office (MTSO). The cell-site controller manages each of the

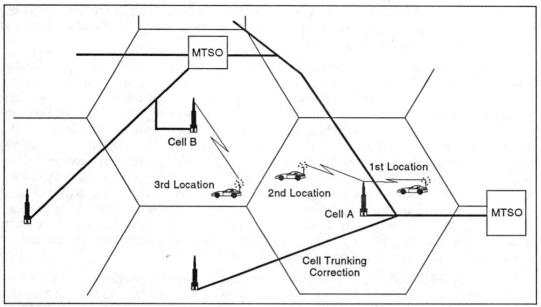

FIGURE 4.2 Cellular Hand-off Process (Constant sampling of power levels indicates when MTSO should switch a cellular user from Cell A to Cell B.)

channels within the cell site. This involves supervising the calls, turning of the radio transmitters on and off and performing operational and diagnostic cells of cell-site equipment. In this respect one of the key roles of the cell controller is carried out through the use of a signaling channel, which is constantly assessing the power and performance of all established links within the cell area. As long as certain performance parameters remain within prescribed limits the controller maintains the call. If at any time the signal should fall below a certain power level, then the controller must decide very quickly where and how to switch the call to another adjacent cell.[5]

In some cases because of the "fuzzy boundaries" that exist between cells, the call can be prematurely switched and create a disturbing ping-pong effect which can lead to a high level of noise, one or more channels of crosstalk or even a dropped call. If one assumes that the boundary in question is reasonably well defined and the antennas in the adjacent cells are operating properly, then the following process takes place. The controller in the cell where the call has been operating polls all of the cells immediately adjacent and then selects the one that indicates through a measurement process the highest power level to the car to the optimum cell. The controller then finds an available frequency and then initiates a complex hand-off processing process. This means that the call is reassigned to the selected new cell site on the next available frequency.[6]

Since the relative power levels represent the only basis for the decision to switch the call, there are a number of opportunities for difficulties to arise. The car may be stopped at the point the switching decision is made and may then, in fact, turn and head in a different direction. Also, the car may be at a high elevation or at a blocked location such as a tall building. A few seconds later a different switching decision again becomes necessary. There is, in short, no predictive capability or artificial intelligence in the decision-making process with regard to the switching of cellular calls.

The process of determining that a call needs to be switched and then actually switching it to a vacant channel in an adjacent cell is a very processing-intensive activity that ties up switching and the cell controller's processing capacity. This means that if there are too many hand-offs of cellular service this leads to system inefficiencies and loss of total system capacity. Critical to the efficiency of a cellular system is the right choice of cellular sizes, the concentration of traffic within those cells, and the degree of cell splitting that is done to respond to heavy zones of traffic usage. The initial cellular systems tended to have large sizes of five miles or nearly eight kilometers in diameter. More recently cell sizes have shrunk to about three miles or about five kilometer diameters in heavier traffic areas. In the case of the new Personal Communications Service (PCS), the size of cells may ultimately be measured in hundreds of yards or meters. Alternatively, however, high-powered or high-tier PCS services offering mobile links to vehicular traffic may still operate cells whose diameters are comparable to conventional cellular services.[7]

The motive that serves to reduce the size of the cells is the need for additional usable frequencies that can be achieved through more intensive reuse techniques. The size of the cells need not be equal in size within a cellular system and may be designed at the outset to have smaller cells in the peak traffic regions of a city. In other cases, it may become necessary to resplit cells because of the build up of traffic in a localized area. In this case, the cell is typically divided into two zones and thus the cellular traffic in that area can be increased, but not doubled, due to inefficiencies that are involved.

There are several factors at work that prevent an actual doubling of capacity. One of the results of splitting the cell is that there are quite logically many more hand-offs that will occur. In the case of PCS with its smaller cells an increase in hand-offs will also naturally occur. The illustration in Figure 4.3 shows that there is an impulse level increase in hand-offs that occur in such a circumstance.[8]

As noted earlier, each hand-off creates the need for intensive processing within the base station systems and the MTSO switch. This serves to limit the capacity of the switch. This process also creates another node to be interconnected and this also limits total system efficiency. In summary, cell splitting, despite its increase in system capacity, is costly and increases congestion in cell switching and controlling systems. It is far better to perform detailed traffic studies and plan the system to match traffic density patterns from the outset, rather than reengineer and reequip the systems through cell-spitting activities. Cell splitting may seem as if it might be a quick fix, but it is a costly quick fix.[9]

The same basic architectural design and hand-off procedures through cell controllers apply to pedestrian-based systems such as DECT, Telepoint, and IMS systems within building complexes. Since decisions by cell controllers involve pedestrian traffic rather than vehicular traffic the speed of decision and processing requirements tend to be less pressured and less subject to some of the confusion indicated above regarding changing traffic patterns and environmental and building interference patterns.

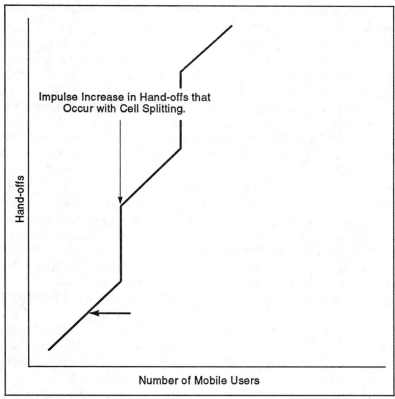

FIGURE 4.3   Inefficiency of Cell Splitting

4.2.2 Transmission and Modulation Systems

For many years the key to worldwide telecommunications systems has been the use of analog-based voice systems running through copper wire. This was the essence of the American Bell system but it also was repeated many times over around the globe. The idea of circuit-switched systems was well entrenched since the service requirement was for telephone connections. Also, analog systems made a good deal of sense to support voice requirements. One system was to create a wave form and modulated it to reproduce a model of speech. This approach was used very extensively in wire-based systems but was also emulated in wireless radio systems when they came into service. The predominant approach that was used for transmitting voice over radio signals was thus to use analog techniques. The most common methods were Amplitude Modulation (AM) and Frequency Modulation (FM). In the FM system the analog model was created by variation of frequency while holding the signal amplitude constant. The other analog method, namely amplitude modulation, was first used for radio transmissions. In this case the amplitude was varied to modulate the signal. The higher-fidelity system was that of FM and it became common for most telecommunications applications. It is still used in several multiplexing systems. The most common analog-based multiplexing system for mobile communications today is called Frequency Division Multiple Access.

Since most applications of the future will involve digital encoding and transmission, the focus will be on the key digital technologies of the future. The key aspects of digital communications involve the conversion of an analog signal to digital format through a process called quantization. This is followed by encoding the signal and modulating it. The completion of this activity is called signal processing. This includes the multiplexing, or the combining of processed signals, and other steps needed to send a signal on a carrier form over a distance. There is then the reversal of the process at the other end of the cycle at the receiver end. This is largely the inverse process of the activities that initiated the process. The end result is a digital-to-analog conversion, which creates what a cellular subscriber would hear as someone's voice.[10]

There are also complexities beyond these basic steps such as processing the signal within the network to "regenerate" it rather than simply receiving a signal and amplifying it along with environmental noise. Signals can be processed in very sophisticated ways. These can include combining digital processing with digital compression techniques to achieve very high rates of information rate transfer per hertz of used spectrum. With advanced digital techniques, performance as measured against this index of hertz bits of information continue to increase steadily. We have progressed from less than half a bit of information per hertz to rates as high as three to four bits of information.

In digital encoding a signal in the form of incoming information can be simply divided into a series of "1's" and "0's" by a complete phase shift or a 180-degree rotation. This bi-phase shift keying of information can encode a signal or information with a high degree of fidelity. One can also use quadra-phase shift keying, which encodes information by using only a 90-degree shift and so on. This involves a trade-off between information density and throughput versus the quality of the signal. More and more information can be very efficiently encoded by using smaller phase shifts, but the chance of error and loss of signal keeps increasing. As one moves from simple biphase shifting to multiphase shifting the information rate per hertz does increase and the representation of this on an oscilloscope is seen as a series of dots of 4, 8, 16, 32, 64, or perhaps even 128. The introduction of only a very small amount of interference or noise, however, can very easily experience a total loss of signal particularly when one moves up to above 16-phase encoding.[11]

These complex phase shifting encoding techniques are thus referred to as multilevel encoding. The addition of only small amounts of noise to the process and the resolution between the different phase shifts is immediately lost. The extremely low noise and bit error rates of fiber optic cable is considered well suited to multiphase shift keying, but the higher noise environment of free space makes this application much more challenging.

The idea of encoding of a digital signal has really two aspects. One is the mathematical concept that information is recorded as a 1 or a 0. In fact, there is actually a physical representation of that information as a pulse or nonpulse. These pulses are never really the perfect notched wave forms that are shown as theoretical models in textbooks. Such a perfectly defined depiction is only the idealized representation. The physical waves can be irregularly shaped enough to be misperceived, particularly if steps in the pulse are taken to represent a 22.5-degree, 45-degree, or even a 90-degree phase shift.

The process of quantization, which starts the process of analog-to-digital conversion or ends it at the remote location, is actually quite simply a voice-created analog wave form that looks like some variation of a single wave. The idea is that there is a multiple step quantization scale that is used to convert the rounded curve of the analog signal into a reasonably close approximation of the original signal. Instead of the signal having an infinite number of smooth steps, the new digital representation would be in error in that it would include any noise already associated with the signal and it would over-represent or under-represent to a small degree most of the sampled wave form.[12]

Nevertheless, the approximation is still very close to the original. Furthermore, once the digital wave form is created, the use of digital transmission techniques makes it largely impervious to the addition of more noise or inaccuracies in the dequantization process in the digital-to-analog conversion.

Once the signal is quantized, then the encoding process is accomplished to convert the information that is the digital model of an analog voice into very simplified form. This results in a string of "1's" and "0's" that contains in compact form all the information needed to recreate the voice signal at the end of the process when the signal is converted back to voice. Once completed, then the next critical step of modulation begins. See Figure 4.4 to see an idealized view of the quantization process.[13]

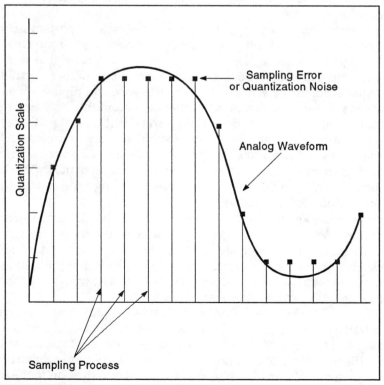

FIGURE 4.4 Analog-to-Digital Conversion

Modulation is simply the variation of a wave form to create a signal that can be transmitted over a distance through some telecommunications media. Since the digital signal is quantized and encoded into simple "1's" and "0's" it may be the thought that this is very simple and that there might be only one or a very few ways to do this. In fact, there are a very large number of analog, digital, and even hybrid digital and analog techniques. There are today, however, two primary digital multiplexing techniques that are strong candidates for digital wireless systems and indeed are in contention with one another. These are, as explained further in Chapter 8 on standards, Time Division Multiple Access (TDMA) and Code Division Multiple Access (CDMA), also known more simply as Spread Spectrum. These digital systems promise expanded efficiency over today's Advanced Mobile Phone Service of three to 15 times. There are, however, proposed new analog systems, particularly Narrow-Advanced Mobile Phone Service (N-AMPS), which itself promises a threefold gain in performance.[14]

Advanced digitally based multiplexing techniques are a major way in which to increase the efficiency of communications systems. The key to measuring the efficiency of a telecommunications system is the amount of information, in the form of voice, data, or images that can be sent through it within a fixed amount of spectrum. Trying to improve this performance measure will often, however, lead to increased power consumption and increased overheads in the system. Before examining the relative merits of TDMA versus CDMA multiplexing systems, it is important to note the general advantage of digital modulation over analog systems.

There are, in fact, many advantages but three of the more important reasons are as follows: (a) the ability to use multiphase encoding in low-noise environments; (b) the ability to use intense signal processing, i.e., digital compression and signal regeneration techniques either to send more information more efficiently or to combat high noise; and (c) the ability to operate successfully within a high-interference and high-frequency reuse environment. These factors will be discussed further below, but the third factor is particularly relevant to modulation techniques. Both TDMA and CDMA, as digitally based multiplexing systems, can operate in a high co-channel interference environment. CDMA by it reliance on unique codes and spreading its signal across a broad spectrum is particularly well suited to operating in an urban environment with a high level of noise. This is true whether one is operating with ultrasmall-aperture antennas accessing a communications satellite or within a digital cellular environment.[15]

As noted earlier, the key to cellular systems is the availability of frequency and its efficient reuse. Today, in analog-based AMPS type systems the reuse patterns involve every seven to 12 cells, although advanced N-AMPS systems have been able to reduce the separation of reuse patterns to the one-in-four or one-in-five level. It is thought that with digital cellular systems of the future the reuse patterns could be as low as every three or four cells. This, of course, results in an enormous efficiency gain.[16]

The idea of Time Division Multiple Access (TDMA) is quite simple. Instead of dividing up a satellite transponder or cellular frequency band by slots represented by segments of frequency, the TDMA approach is to create time slots allocated to different users. This approach allows the full power of the transmitting antenna to support a short burst of signals to be sent and then quickly transition to the next burst and then on to the next. Although we tend to think of voice, data, and video services as being continuous in nature, in a digital system this is really not so. The "1's" and "0's" are actual discrete and discontinuous events. The code that represents voice or data or video can indeed be sent in bursts measured in milliseconds and processed together at the other end of the TDMA transmission. Despite the discontinuous or discrete nature of the service, it still creates the appearance of a continuous service. The separation of the carrier in the time domain is very efficient for either satellites with multiple beams to interconnect or for the various cellular systems where there are again many different beams to link together. TDMA systems based upon the well-established Electronic Industries Association's IS-55 standard promise gains of three to five times. There is a new more advanced approach called Extended-TDMA, developed by Hughes Network Systems that allows further optimization and compression so as to increase capacities up to 15 times over conventional AMPS service. All TDMA systems based upon the IS-55 standard require user authentication and encryption to protect user privacy and system integrity. From a technical perspective the same performance levels and characteristics likewise apply to the GSM cellular systems, which are also based on GSM.[17]

The idea of Code Division Multiple Access (CDMA) was first developed within the military communications systems of the United States. The motivation was, in fact, to develop a highly efficient communications system that was heavily encrypted, difficult to jam, and highly robust even in a high-interference environment. It turned out that CDMA was very proficient in all three regards. Its reliance on heavy processing power to extract a coded signal embedded across a broad frequency spectrum was almost hidden among the various signals against which it was overlaid. The only way to extract the wanted signal from among many other unwanted signals was to have the right code. Furthermore, it was already digitally encoded so that it was in effect already encrypted.

CDMA was, however, thus not only good for military purposes but also well suited to commercial communications requirements as well. Thus, the heavy coding served to protect the signal against jamming on one hand but to also protect it against a heavy noise environment in an urban environment. Further, it was also suited to work in cellular systems with more intensive levels of frequency reuse. The use of coding allows more channels to be derived by the overlaying of carriers one over another. In the case of CDMA and TDMA there are certainly higher overheads and more power consumed than in analog systems, but the net efficiency gains over analog systems clearly justify these trade-offs. Within ten years virtually all wireless systems and satellite systems will have converted from analog-based multiplexing systems such as FDMA and will have instead migrated to either TDMA or CDMA.

The diversity of standards for mobile telephone systems has become an increasing problem in its own right. The answer has been the IS-41 standard of the EIA, which defines the signaling procedures and hand-off rules to interconnect disparate systems such as AMPS, TDMA, and in the near-term future, CDMA.[18]

4.2.3 Signal Processing and Regeneration

The intersection of telecommunications and computer science took place in a real and tangible way over a decade ago. In fact, if you count the telegraph as a digital instrument, the historical relationship can be counted back over a century. The basic idea is that the key functions of telecommunications are increasingly a specific form of digital processing and that the tremendous gains being achieved in advanced processing and software development when applied to telecommunications offer major gains in performance and in cost efficiency. Furthermore, the tools of telecommunications such as signaling equipment, switches, and radio transmitters are increasingly resembling computers with very specialized software.

This general trend toward universal digitalization, of course, has a few exceptions, but the bottom line is that telecommunications and computer processing are today among the world's fastest growing industries. These technologies are experiencing the fastest rate of productivity gain (i. e., Moore's Law), and are among the most cost-efficient and profitable enterprises on the globe. This parallel growth and development should not really be surprising since the convergence process has made digital processing and telecommunications very closely interlinked indeed.

It is in the area of signal processing that this parallelism is perhaps most clearly apparent. A digital signal processor (DSP) is, in fact, a computer that is software defined to perform specialized telecommunications functions. These functions include using generic or proprietary encoding and modulation techniques to compress and to multiplex signals for high-efficiency transmission. DSP can also be used to regenerate signals at the end of the transmission–reception cycle. This process of regeneration can happen at two levels. One is to use processing power to regenerate a digital carrier wave at what might be called the macrolevel. This provides a clear advantage and is a function that a reasonably fast processor can do quickly and efficiently without introducing a great deal of delay or latency in the end-to-end transmission path. The typical advantage in such carrier wave processing is 1–3 db.

The second, much more demanding approach to signal processing is what is called bit-by-bit processing. This means that the entire signal at the receive end (or on-board a satellite in the frequency translation and retransmission process) is completely reprocessed with each bit of information being totally regenerated to create a "perfect" new source transmission. This requires a huge increase in processing power and unless a very-high-speed processor (i.e., a super computer) is utilized this can introduce a very high level of delay in the transmission process. If one does have the software and a super computer available, the gains in very-long-distance transmission such as in the case of a geosynchronous satellite link can be very impressive (i.e., 6–9 db advantage) or 5–6 db for a medium earth orbit satellite.[19]

The subject of digital compression techniques will be addressed in detail in the next section, but it is important to note here that the various forms of digital signal processing, which includes multiplexing and advanced encoding, signal regeneration to create a new "source," and digital compression combine to contribute enormous efficiency gains. These collective advantages, which allow new digital wireless systems to be ten to 20 times more efficient than conventional FDMA wireless signals, are indeed the most important gains in performance of all the new systems that will be addressed in this chapter. Digital signal processing is thus the central technology in all wireless telecommunications efficiency advances today and a number of years into the future.[20]

4.2.4 Digital Compression

The single biggest factor that is currently advancing wireless telecommunications within the broad technologies that represent DSP is digital compression. It was thus seen useful to have at least a subsection devoted to this subject.

First of all, there are two basic approaches to digital compression that are typically carried out by the processors and coders/decoders (codec) located within the multiplexer. These two activities are digital speech interpolation (DSI), which obviously relates to voice services, and the other is interframe or intraframe digital compression, which can be applied to all services. Digital compression reduces the information rate required to send a service through a telecommunications system by applying complex algorithms that somehow reduce the amount of information required to complete the information exchange transaction. This could be video or imaging wherein a high degree of compression is easiest to achieve or it could also be voice, or even data, where compression is most difficult to achieve.

The algorithms involving video compression are often based upon assumptions of stability of image over much of the display screen with updated information concentrating on that part of the image that changes frame to frame. In applications such as video conferences or newscasts this technique can work very well, but in cases such as a car race or a dance contest with a great deal of motion and image changes this is much less successful. Other algorithms are based upon numbering in a code book many different complex patterns of pixels with a processor matching patterns of pixels to the closest code book equivalent.

Over the last decade the algorithms that allow digitally compressed telecommunications to become more effective have constantly improved. These techniques have allowed broadcast images to be created at a level as low as 6 megabits per second in contrast to the former rate of 68 megabits per second. These systems are now fully operational in the Hughes DirecTV satellite system, for instance. Video conferencing has dropped from 3 to 6 megabits per second down to rates in the 64–384 kilobits per second range. Different algorithms have allowed voice transmissions to be achieved at slower and slower speeds (i.e., from 64 kilobits per second down to 4.8 kilobits per second). Even in the area of data transmissions, improvements of 20–30 percent have been achieved.[21]

The key to advances in this field are today based upon the use of artificial intelligence for predictive patterns to enhance compression gains. Improved performance is also being achieved through the creation of very sophisticated, processor-intensive algorithms that produce very good analogies to the originals with much, much less information. In the past the cost of such processors in codec equipment would have been much too expensive and would have introduced much too much delay to have been commercially feasible. Thus the speed of processors continues to increase while the cost falls. This means that the feasibility of even better performance in the area of digital compression increases even as codec costs seem to fall rapidly. This continuing benefit to telecommunications as a result of Moore's Law seems to be particularly evident here, as can be seen in Figure 4.5. (Moore's Law incidentally predicts a doubling of performance every 18 months.)[22]

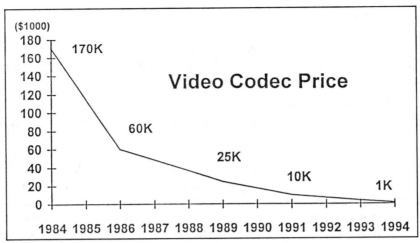

FIGURE 4.5   Image Technology Trends

4.2.5 Key Technical Constraints and Overcoming Them

The key technical elements in the field of wireless telecommunications are actually reasonably few in number and quite straightforward. These are digital processing, advanced frequency reuse techniques, advanced modulation, advanced signaling, and network control and management techniques.

Digital processing is being utilized more and more to gain an advantage in a wide spectrum of areas. These digital applications can be categorized in the following ways: (a) digital compression and digital speech interpolation; (b) signal regeneration; (c) digitally created or software-defined telecommunications equipment; (d) miniaturization of terminal devices through advanced DSP microprocessors design; and (e) digital processing to promote frequency reuse (i.e., more tolerance to noise and co-channel interference and more effective beam interconnection). The combined force of these uses of DSP has moved the cellular/PCS industry ahead rapidly in the last few years with another order of magnitude of performance increases yet to come.

Advanced modulation techniques coupled with error-control systems can serve to send information more securely, with high-quality standards (i.e., low bit-error rates), and with higher throughput rates. The leading modulation systems for wireless systems today are CDMA and TDMA—each with special advantages and liabilities. It is of interest to note that certain new modulation techniques that are under development as optimized systems for fiber optic cable, such as so-called "color multiplexing," or more formally, Dense Wave Division Multiplexing, are really not compatible with either CDMA or TDMA. This could eventually mean expensive and delay-inducing remodulation systems to interconnect fiber, wireless, or satellite systems.

Another key area of technological development is that of advanced signaling systems. If a modern digital telecommunications system were to be considered to be analogous to a transportation system, then the highways and railroad tracks would represent the broad band transmission facilities equivalent to fiber optic cables, satellites, and microwave relays. The local streets would represent copper wire and cellular communications transmission towers. The switching network would be represented by clover leaf interchanges and traffic signals at intersections. The signaling system, which is the intelligence in the network, would be the traffic cops, the transportation planners, and the navigators within the cars, train, and planes. They control the routing of calls and much, much more. The artificially intelligent signaling system today can provide automatic number identification, verification of line accessibility prior to connection of the bearer or voice channels themselves, call forwarding, multiple ringing patterns, and so on.[23]

The power of this "smart" signaling or network management system is significant. By more efficient routing of calls and by placing calls on the system only when they can be connected greatly expands the capacity of the network. The ability to reroute calls flexibly in case of system failures or the ability to connect roaming customers directly all serve to reduce hardware investments by substituting more cost-effective software instead. Simply put, the low cost of processing power and "smart" network software can make existing telecommunications networks of all types, wire and wireless, much more efficient and reliable.

In particular, the introduction of Signaling System Number Seven technology allows expanded network control and management, greater flexibility to handle a great diversity of services, and aid network restoration in contingency situations. Although signaling technology is much different from the other types of technology discussed in this section, it is nevertheless quite important. The creation of enhanced features and value-added services via the new "D" or signaling channels within the ISDN services is clearly a key step for the world of telecommunications. The fact that most cellular telecommunications have a throughput capacity below that of basic-rate ISDN (i.e., 144 kilobits per second or even well below the basic rate D channel at 16 kilobits per second) is seen as a problem. The primary rate ISDN at 1.544 Megabits per second with a "D" signaling channel of 64 kilobits per second is, of course, well beyond the capacities of today's mobile systems below 2 GHz. In short, today's cellular mobile services are crammed within frequency allocations that are inadequate for the service requirements or the associated signaling. In the future, when more wide band allocations are made at much higher frequencies, it would be hoped that the lack of signaling channel capacity will be effectively addressed so that mobile services cannot only be of high quality but increasingly intelligent and capable of enhanced services as well.[24]

4.3 IMPLEMENTING MOBILE COMMUNICATIONS TECHNOLOGY

The amazingly rapid growth of wireless technology in the last decade is truly impressive. Ten years ago there was only a start-up virtually nonexistent cellular industry using crude FDMA analog-based technology to provide a poor-quality voice service at fairly high prices. Today we have moved to an increasingly sophisticated cellular industry that is moving rapidly toward digital and advanced analog systems featuring value-added data (CDPD), fax, e-mail, and integrated paging services. The idea of a mobile virtual office is, in fact, being realized. We have progressed from simple bent-pipe satellites to those with on-board switching and processing capabilities. We have moved from conventional microwave to digital microwave. We have seen the growth of a wide range of wireless services that now represent a $100 billion market and that will likely be double that amount by the early twenty-first century.

Despite the rapid growth of this technology, the key is still in the effective application of these technical developments. Without certain enabling steps the technology will either remain unutilized or underutilized. These issues of technological implementation within the wireless industry include: (a) accessing new or better utilized frequency bands; (b) ensuring competitiveness and/or compatibility with terrestrial technologies; (c) implementing the most effective operational and management systems; (d) avoiding key pitfalls with regard to standards issues; (e) considering issues of health and environment with regard to system planning and operations; and (f) addressing the issue of how technological planning and implementation is affected by the key trend of the globalism of telecommunications. These six issues are key areas wherein the creation of new wireless technologies can be ensnared in a number of difficult and sometimes rather unexpected problems. Anyone planning to implement a wireless network should probably consider all of these points carefully.

4.3.1 Getting Access to Spectra

Clearly obtaining access to spectrum is a critical aspect in the successful operation of any wireless system. There are really only a limited number of ways in which this might be done. One is, of course, to formally file to seek a license, whether this be for a radio or television station, an IMTS type service, a SMR or cellular service, a microwave relay, or even a new satellite network. The problem with this approach is that there are only a limited set of frequencies available for various services and in most cases those licenses are already assigned to others. Only when new services such as was the case with cellular in the mid-1980s, or low earth orbit satellite systems in the early 1990s, or PCS in 1994, are new opportunities opened up in this manner. Even so, this process can be long and time consuming. The timing between applying for a license in a new wireless service and actually receiving a license, at least in the United States, is typically over two years and can be much longer. Similar delays can be anticipated in many other countries such as those in Europe, and Canada and Japan as well.[25]

Another strategy that has been used successfully in the past is to buy out someone else who is providing service in the market you wish to enter. A key variation on that theme is represented by the Enhanced Specialized Mobile Radio industry. Here the strategy has been to buy out a user of a frequency who has been applying it in an inefficient way and then to apply modern technology to derive much higher capacity and value from the spectrum than in the past. This is exactly what has happened with the sale of so-called SMR licenses. The buyers subsequently converted these frequencies to ESMR service through the application of digital cellular techniques to the old dispatching service frequencies. Yet another variation on this theme is the possibility of joining in an enterprise that is acquiring access to new or reconverted frequency. This is represented by the case of Motorola, which acquired about one-sixth of NEXTEL by trading in its extensive SMR licenses on the understanding that they would, in effect, be converted to higher-value ESMR licenses.[26]

Finally, there is the simple strategy of reengineering the use of the frequency already available under existing licenses. This can be accomplished by use of digital compression techniques, by use of improved modulation techniques, by redivision of existing cells in a mobile system, by adding digital microwave to existing microwave systems, and so on. In general, operators of wireless systems should measure the number of channels or derived capacity achieved per available megahertz of capacity and try to improve upon these results as an efficiency measure. Such a figure of merit is very helpful in plotting a strategy of modernization and system expansion. With the latest in encoding and modulation schemes one cannot only use smaller and better defined cells, but also reuse the frequencies more intensively (i.e., every three to four cells rather than one in six). The basic fact is that the intensity of use of frequencies for cellular as well as satellite is constantly going up and the value of frequencies through the auction process in particular is going up even faster. One can no longer afford to stand still in terms of ongoing innovations in wireless telecommunications technology.[27]

The number one problem with wireless mobile communications implementation today is quite simply obtaining access to frequencies through some form of licensing process. The difficulties already noted in this respect are in fact driving some to move sharply toward the unlicensed wireless market and have customers use these readily accessible bands that are available to all who wish to operate in these frequencies on a noninterference, low-powered operation basis. The so-called Industrial and Medical Services band known as Part 15 under the existing FCC regulations and the new 30 MHz of unlicensed PCS bandwidth makes this an attractive option to those unwilling or unable to bid many millions of dollars for spectra.[28]

4.3.2 Cable versus Wireless Communications Systems

The most unproductive question to ask concerning telecommunications is whether fiber optic cable or wireless telecommunications is the best technology to use. Clearly, the best approach is to utilize the strengths of both technologies. This means in the case of mobile systems to transmit messages to and from vehicles by radio waves and to use cable technology to create node interconnectivity. Part of the problem in sorting out the roles and respective capabilities of wireline and wireless systems is that many misunderstandings exist. The basic misconception is that fiber has a virtually infinite band-width and should be used for all broad band requirements, while wireless systems have only narrow frequencies available to them and should probably be best used for narrow band mobile services. In fact, with modern cellular reuse techniques, digital modulation and encoding techniques, and increasing exploitation of the EHF bands, there are no fundamental constraints on wireless broad band delivery systems. What this means is that it is possible to conceive of satellite systems or wireless networks in the millimeter wave band that are capable of 10–100 gigabits per second in throughput. Such systems could carry all services, including a super computer interconnect.

The similarity and differences between fiber optic cable and satellites are rather generally misunderstood. The current comparison of the two media in terms of key performance measures is provided in Table 4.1.[29]

	Advanced Satellite	**Advanced Fiber Optic Cable**
System Availability	99.98%	99.98%
Bit Error Rate (BER)	10^{-7}–10^{-11}	10^{-7}–10^{-11}
Capacity/Bits/seconds	1–3.2 Gb/s	840 Mb/s–2.5 Gb/s
Transmission Delay	270 ms	Under 50 ms
Typical end-to-end network transmission time	350–1000 ms	200–1400 ms

TABLE 4.1 SATELLITE VERSUS FIBER OPTIC CABLE

Likewise, it is possible to envision cellular systems that can carry not only data and highly compressed voice, but also video and imaging services. Such systems would, of course, require very broad band frequency allocations, most likely in the millimeter wave band. For intraoffice and intrahome wide band services, the frequency of choice may actually be infrared communications, where the spectrum is virtually unlimited.

The new technologies, especially highly efficient cellular reuse techniques, have thus rewritten the rules. Wire and wireless systems combined can allow us to achieve almost impossible feats of information relay and sharing. Mobility and wide band can and will go together in the future. New architectures that are more than just concentrated traffic through switched hierarchies can be designed to allow mesh-like connection of any point to any point in a network. The key to these networks will more likely be related to the speed of processors than the transmission mode selected, whether it be fiber optic cable or wireless communications.[30]

4.3.3 Designing Operating Systems for the Future

It is often the case that telecommunications planners get so involved in designing transmission and switching systems that they neglect the important points related to operational reliability, redundancy, and operating costs and procedures. The highest capacity network design is not necessarily the most reliable, restorable, or cost efficient. Most large businesses do not make their decision based on the capital costs to implement a new system. In fact, the criteria include such considerations as: (a) system availability; (b) quality of service; (c) cost to operate and maintain; (d) restorability; and (e) upgradability and modular expansion capabilities. This is a general set of criteria that might be easily applied to a wireless LAN or PABX system, a private ESMR service for a corporation, or a dedicated satellite network.

A private customer selecting a cellular radio and "virtual office" setup might equally well consider most of the above points plus the additional issues of convenience and value-added features of portable wireless terminal equipment. It is thus important to know how well technology works and how reliable and cost effective it is in its longer-term operation.[31]

In the case of PCS service, for instance, the number of node interconnections in the system compared to that of a conventional cellular system might involve an exponential increase in nodes. This would be particularly true if so-called microcells of several hundred meter diameters were to be used. This could easily mean more complicated processing, signal routing, switching, and billing data record keeping. In some systems billing operations and collections involve comparable expenses to those involved with the networks operation. Ultimately, in modern wireless telecommunications operations, one must seek systems that are highly automated, artificially intelligent, self-restoring, capable of upgrading to provide new value-added services, and able to capture and instantly display billing information.[32]

The issue of network management for wireless telecommunications is also closely tied in here as well. The integration of network management, automated billing, self-restoration of service, value-added services, and competitive advantage in the marketplace are today closely interrelated. The creation of network management systems that just allow a network to run is no longer sufficient. All of the critical functions cited above must also be provided in the current world market. This is true of wire and wireless services alike.[33]

4.3.4 The Pitfalls of Standards

The concept of a technical standard whereby universal agreement is reached to provide a product or service in a single and precisely agreeable way seems to have a number of points in its favor. Standards allow for economies of scale and scope to apply. They allow multiple suppliers to compete, but they also assure consumers of interoperability and compatibility when they purchase a product or a service. Nevertheless, standards can also create problems as well. Once a standard is set, there tends to be a block to innovation and acceptance of new technology. Further, in today's complex world the creation of a new standard in telecommunications, in general, and in wireless telecommunications, in particular, is often dominated by a few large commercial organizations. A small, innovative, start-up company with a very good technology, short of teaming with a giant, may have great difficulty of having their technology accepted as a new standard.[34]

There are many examples of smaller companies and how they have attempted to enter the wireless technology market and cope with the standards issue. In the case of CDMA modulation, there was example of the Equatorial Communications Company, started by Dr. Edwin Parker of Stanford University. Eventually, this innovative enterprise, which sought to harness spread spectrum technology for rural and remote satellite education was taken over by Contel/GTE Corporation. This was essentially a "forced" solution due to marketing and standards issues.[35]

Qualcomm, Inc., the leader in CDMA terrestrial mobile communications, has attempted a wide range of strategies to fulfill its goals of growth within an uncertain standards environment. This has included such steps as licensing of the OMNITRACS service to EUTELSAT under the trade name EUTELTRACS. They have also joined with ALCATEL/Loral Corporations in the launching of the Globalstar satellite system to seek a broader scope of financial and marketing support as the status of CDMA in the standards-making process remains in contention. In short, the extremely strong intellectual position presented by Qualcomm's two engineering pioneers, Dr. Viterbi and Dr. Jacobs, is indeed impressive. It is nevertheless not necessarily enough to convert technology and interesting applications into a global standard without a lot of support from either national governments or very large corporations. This is to say that standards-setting is ultimately a political process and the best technology may not always win out in this process.[36]

The mechanisms to address this process of converting a new and innovative technology into a standard are severalfold and are worthy of note. These mechanisms include the following:

(a) **Pioneer's preference:** A developer of a truly important new technology can apply to the FCC for a pioneer's preference. This approach, if successful, bestows the benefit of a frequency award if a wireless service is involved and a six-month head start over would-be competitors. If the application is unsuccessful, such as Motorola's application on behalf of the Iridium Satellite System, then this can be a liability. The applicant has had the expense of the application, the disclosure of vital technical and operational information, and no real gain at all. In the case of Omnipoint, this helped to win an early PCS license, but considerable costs still applied to the early license.[37]

(b) **Teaming:** A smaller start-up firm might team with one or more larger and more powerful entities though licensing of the technology, stock exchange or other partnership agreements, or simply selling out to someone with much greater financial, marketing, and policy influence. In some cases, this may not be a matter of standards but a need to validate one's new method of market entry. Such an example is the aggressive move of NEXTEL into the new Enhanced Specialized Mobile Radio service whereby their partnership with Motorola has provided credibility and market influence. This is, of course, not necessarily enough. Ultimately it will be the quality of the service that decides market success rather than which partner is involved in the enterprise.[38]

(c) **Intellectual Property Rights:** This is a less certain way to proceed in an international wireless telecommunications market. The case of CDMA implementation in the mobile communications field reflects the dilemma rather well. It is today very hard to protect advanced encoding and modulation techniques because there are often parallel and comparable ways to achieve similar results without direct duplication. Proof of duplication in software and conceptual matters is often difficult and if the issue is considered in a potentially unfriendly international venue, the outcome may not be favorable.[39]

Even if one is successful, then the patent or licensing fee may very well serve as a barrier to the technology being accepted and certainly a barrier to its acceptance as an international standard.

4.3.5 Global versus National Strategies

The United States has been a major leader in implementing wireless telecommunications in almost every aspect, i.e., broadcasting radio and television, microwave relay, satellite communications, AMPS cellular radio, and now PCS services. Wireless services and the supply of wireless equipment products are both growing very dynamically around the world. In terms of wireless telecommunications technology and the worldwide supply of equipment, it is accurate to say today that Japanese industry is now predominant at least in terms of terminal equipment.

The top suppliers of wireless transceivers in the U.S. market today include: Panasonic, Audiotel/Toshiba, Oki, and Novatel (a Canadian, Hong Kong, Korean, Taiwanese company). Motorola is the only U.S. supplier in the top ten. In the world of satellites, U.S. suppliers such as Hughes, Loral, Lockheed Martin , TRW, CTA, and so on, are still predominant but a recent study conducted by NASA and the National Science Foundation concluded that research and development in Japan as well as in Europe would close the gap vis à vis the United States within the decade.[39]

There should be no presumption of a U.S. technological lead in wireless telecommunications. The United States is the largest single world market, but its technology is at best competitive with that of Japan and Europe—not more advanced. The ability to have a large home market or to define a technical standard no longer ensures market success. One must recognize that a manufacturer of wireless telecommunications equipment or a provider of such services must be able to compete on a global level. The meaning of "home markets" is increasingly more difficult to define in a world of open trade, regional trade organizations (e.g., EEC, NAFTA, etc.), and international partnerships like Novatel that cross many boundaries. This problem is likely to become more and more common.

The future of global trade seems headed toward more open competition, more mergers and partnerships, and wider trade coalitions. Wireless telecommunications can only be expected to become more international and even global in scope. This is because it is a high-growth market sector, and it also exploits the latest in high technology. This strong globalization trend is also likely to be promoted by complex international technical standards. Finally, globalization is definitely the result of numerous new mergers and acquisitions that are increasingly reaching across international and even regional boundaries.[41]

4.4 INVENTORYING KEY PROBLEMS AND THEIR SOLUTIONS

The speed and sophistication of technological developments in the field of wireless telecommunications technology is truly amazing. It seems quite hard to believe that IMTS technology with no effective reuse of frequencies was the norm for mobile communications just a little over a decade ago. It seems remarkable that mobile users worldwide are moving up toward the level of 20 million and that with new PCS services, ESMR, and digital cellular options available, usage levels may exceed 100 million in another decade.

Equally optimistic forecasts for wireless LANs and PABXs systems have been projected from current trends. Here, the desire for flexibility, reconfigurability, and quick access is also leading to a rethinking of what an office environment should look like in the twenty-first century. Digital innovations should spur the growth in all these areas by offering more cost-effective service and opening up a wide range of new services.

Other terrestrial wireless telecommunications sectors in the broadcasting, microwave relay, and emergency and safety services sectors will likely not experience rapid growth, but still pose interesting challenges in terms of technical innovation and new operational implementation concepts.

In light of the general trend of rapid service growth and effective development of new technology, it would seem that a very favorable assessment could be given to this field. This is, however, not to say that there are no problems to consider. There are at least eight key problems of technology and operational implementation to overcome. These can be quickly summarized as follows:

(a) Avoiding or overcoming problems of inconsistent allocation of frequencies for mobile or other wireless services in different parts of the world.

(b) Avoiding or overcoming problems of inconsistent or incompatible standards with regard to signaling protocols, channelization, modulation, encoding, and multiplexing techniques, and so on.

(c) Achieving compatibility of interconnection with fiber optic systems in terms of error-control techniques, storage and buffering, and modulation, encoding, and multiplexing techniques.

(d) Achieving compatibility of interconnection with satellite and High Altitude Long Endurance (HALE) platforms, especially with regard to seamless mobile telecommunications services.

(e) Developing systematic standards for digital compression and digital speech interpolation techniques that allow global seamless services.

(f) Developing standards and systems architecture needed to allow more effective global roaming in the mobile telecommunications sector (e.g., Subscriber Identification Modules).

(g) Creating more versatile and capable wireless mobile terminals for broader band user requirements in such areas as tele-education, tele-health, remote governmental services, and so on. (These might operate to terrestrial wireless as well as higher-powered satellite systems in rural and remote areas.)

(h) Creating new types of wireless applications such as electronic newspapers, navigational systems for "intelligent" highways, public safety and rescue systems, or rescue and other emergency operations.

ENDNOTES

(1) MacDonald, V.H., "The Cellular Concept," *The Bell System Technical Journal*, January 1979, Vol. 58, No. 1, pp. 12-22.

(2) ——, "The Cellular Concept," *The Bell System Technical Journal*, January 1979, Vol. 58, No. 1, pp. 30-32.

(3) Calhoun, G., *Digital Wireless Communications* (Norwood, MA: Artech House, 1988), pp. 97-98.

(4) ——, *Digital Wireless Communications* (Norwood, MA: Artech House, 1988), pp. 102-105.

(5) Whitehead, J., "Cellular System Design: An Emerging Engineering Discipline," *IEEE Communications Magazine,* February 1986, Vol. 24, No. 2, pp. 10-12.

(6) Cooper, M., "Cellular Does Work If the System Is Designed Properly," *Personal Communications,* June 1985; also see Lee, W.C., "How to Evaluate Digital Cellular Systems," Federal Communications Commission, Washington, DC, September 1988.

(7) PCS article.

(8) Calhoun, G., *Digital Wireless Communications* (Norwood, MA: Artech House, 1988), pp. 105-106.

(9) ——, *Digital Wireless Communications* (Norwood, MA: Artech House, 1988), pp. 106-108.

(10) Bellamy, J.C., *Digital Telephony* (New York: Wiley, 1982), pp. 70-78.

(11) ——, *Digital Telephony* (New York: Wiley, 1982), pp. 80-82.

(12) ——, *Digital Telephony* (New York: Wiley, 1982), pp. 90-93.

(13) Calhoun, G., *Digital Wireless Communications* (Norwood, MA: Artech House, 1988), p. 159.

(14) Berthoumieux, D., and M. Mouly, "Spectrum Efficiency Evaluation Methods," *Proceedings of the International Conference on Digital Land Mobile Radio Communications,* Venice, Italy, July 1987.

(15) "Cellular Network Equipment Overview," *Data Pro Communications Series* (New York: McGraw-Hill, 1993), pp. 13-15.

(16) "Cellular Network Equipment Overview," *Data Pro Communications Series* (New York: McGraw-Hill, 1993), pp. 16-18.

(17) "Cellular Network Equipment Overview," *Data Pro Communications Series* (New York: McGraw-Hill, 1993), pp. 18-19.

(18) "Cellular Network Equipment Overview," *Data Pro Communications Series* (New York: McGraw-Hill, 1993), p. 20.

(19) Iida, T., "Advanced Satellites Design and Bit-by-Bit Processing," Pacific Telecommunications Conference, Honolulu, Hawaii, January 1993, pp. 1-25.

(20) "Advanced Digital Signaling Concepts and Issues," NASA Contract Report ITP 92-2 (Interdisciplinary Telecommunications Program, University of Colorado at Boulder, 1992), pp. 112-132.

(21) Wu, W.W., "Advanced in Digital Processing," IEEE Special Forum on DSP, Washington, DC, January 1994, pp. 200-232.

(22) "Trends in Video Codec Technology and Performance," *Northern Telecommunications Technical Report,* April 1992, pp. 8-14.

(23) Bellamy, J.C., *Digital Telephony* (New York: Wiley, 1982), pp. 34-45; also see "The Information Wave: Digital Compression is Expanding the Definition of Modern Telecommunications," *Uplink* (Los Angeles: Hughes Aircraft Co., Spring 1994), pp. 4-7.

(24) "Advanced Digital Signaling Concepts and Issues," NASA Contract Report ITP 92-2 (Interdisciplinary Telecommunications Program, University of Colorado at Boulder, 1992), pp. 32-34.

(25) Manuta, L., "Riding the Spectrum Wave," *Satellite Communications,* July 1994, pp. 24-25.

(26) Schnee, V., "NEXTELS's ESMR Arrives: A New Era of Cellular Integrated Services," *Wireless,* January/February 1994, Vol. 3, No. 1, pp. 17-20.

(27) "Value of Spectrum Keeps Increasing In Expanding Wireless Market," *Inside Wireless Reports,* January 1994, pp. 8-10.

(28) Holland, B., "The Unlicensed Wireless Telecommunications Market" (Broomfield, CO: Spectrallink, 1994).

(29) Carraway, R.L., J.M. Cummins, and J.R. Freeland, "The Relative Efficiency of Satellites and Fiber-Optic Cables in Multipoint Networks," *Journal of Space Communications* (Amsterdam, Netherlands: IOS Press, January 1989, Vol. 6, No. 4, pp. 277-289.

(30) ——, "The Relative Efficiency of Satellites and Fiber-Optic Cables in Multipoint Networks," *Journal of Space Communications* (Amsterdam, Netherlands: IOS Press, January 1989, Vol. 6, No. 4, pp. 292-293.

(31) Kachnar, M., "The Virtual Office," *Wireless*, September 1993, pp. 24-27.

(32) Hardy, T., "Personal Communications Services," *IEEE Communications Magazine*, June 1992, pp. 22-25.

(33) ——, "Personal Communications Services," *IEEE Communications Magazine*, June 1992, pp. 26-28.

(34) Bloomfield, R., "Contemporary Telecommunications Standards-Making Activities," University of Colorado, Boulder, CO, July 1994.

(35) Pelton, J.N., *Future Talk* (Boulder, CO: Cross Communications, 1991).

(36) Channing, I., and R. Burr, "Worldwide Digital Cellular," *Mobile Communications International*, Winter, 1994.

(37) Patton-Foster, K., "The Effectivenss of the Pioneer Perference Process," unpublished Masters Thesis (Interdisciplinary Telecommunications Program, University of Colorado at Boulder, December 1994).

(38) Schnee, V., "NEXTELS's ESMR Arrives: A New Era of Cellular Integrated Services," *Wireless*, January/February 1994, Vol. 3, No. 1, pp. 18-20.

(39) Pipe, R., "Intellectual Property Rights and Telecommunications," *Transnational Data Report*, February 1994.

(40) NASA/NSF, *Panel Report on Satellite Communications Systems andTechnology* (Baltimore, MD: International Technology Research Institute, July 1993).

(41) Wimmer, K., and B. Jones, "Global Development of PCS," *IEEE Communications Magazine* (August 1994), pp. 22-29; also see "World's Top 10 Markets for Telecommunications Investment Forecasted," *Global Telecom Report*, April 4, 1994, Vol. 4, No. 7, pp. 1-2.

CHAPTER FIVE

SATELLITE COMMUNICATIONS TECHNOLOGY AND ITS IMPLEMENTATION

5.1 INTRODUCTION

The technical evolution of satellite communications can probably be seen as having had several distinct phases. The first phase was that of the satellite as a totally passive reflector of communications with no frequency translation, amplification, or switching. Experiments were, in fact, first conducted using the moon as a distant mirror to reflect signals back to the earth. During the early era of artificial satellites, experiments were carried out in the early 1960s with the metallic balloon satellite known as ECHO. The very low usable capacity that could be derived from this technique quickly led to the abandonment of this approach. Similar experiments with the use of meteor showers to reflect signals have also been undertaken with similar limited results.

The next phase of the evolution of communications satellites was the development of active transponders, which translated up-link and down-link frequencies and amplified the incoming signal for retransmission back to the earth's surface. This has been the predominant form of satellite communications for the last two decades. Quite recently, the third phase of satellite communications began to evolve with the new concept that satellites could do more than just relay signals. Instead, communications satellites can now be designed to switch signals and through on-board processing to regenerate the total signal to achieve improved signal quality, route signals more efficiently to multiple down-link beams, improve channel capacity, and allow the use of smaller and less costly ground antennas. The evolution in terms of size, mass, complexity, capacity, and overall performance characteristics for "conventional" geosynchronous satellite systems is shown in Table 5.1. It should be noted that in the 1990s new types of satellite architectures involving lower earth orbits began evolving. These satellites, which require the deployment of many more but often smaller and less costly satellites, follow different cost and performance patterns. A comparison of the various types of systems against a common index is provided later in this chapter in Table 5.2.

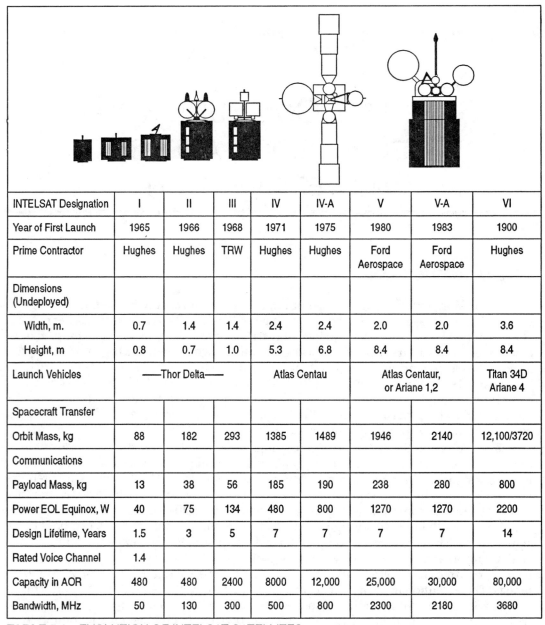

INTELSAT Designation	I	II	III	IV	IV-A	V	V-A	VI
Year of First Launch	1965	1966	1968	1971	1975	1980	1983	1900
Prime Contractor	Hughes	Hughes	TRW	Hughes	Hughes	Ford Aerospace	Ford Aerospace	Hughes
Dimensions (Undeployed)								
Width, m.	0.7	1.4	1.4	2.4	2.4	2.0	2.0	3.6
Height, m	0.8	0.7	1.0	5.3	6.8	8.4	8.4	8.4
Launch Vehicles	——Thor Delta——			Atlas Centau		Atlas Centaur, or Ariane 1,2		Titan 34D Ariane 4
Spacecraft Transfer								
Orbit Mass, kg	88	182	293	1385	1489	1946	2140	12,100/3720
Communications								
Payload Mass, kg	13	38	56	185	190	238	280	800
Power EOL Equinox, W	40	75	134	480	800	1270	1270	2200
Design Lifetime, Years	1.5	3	5	7	7	7	7	14
Rated Voice Channel	1.4							
Capacity in AOR	480	480	2400	8000	12,000	25,000	30,000	80,000
Bandwidth, MHz	50	130	300	500	800	2300	2180	3680

TABLE 5.1   EVOLUTION OF INTELSAT SATELLITES

Thus, simultaneous to the evolution of the "smart" satellite with switching and regenerative signals we have also seen the evolution of different types of satellite architectures. The most notable of these have included plans for the deployment of satellites in different types of orbits. Today there are many different types of orbital configurations, such as geosynchronous satellites, medium earth orbit satellites, low earth orbit satellites, and elliptical orbit satellites. These satellite configurations can be variously used to provide voice, video, data, radio, position determination, direct broadcast, mobile, continuous, or store and forward messaging services.[1]

As the nature, characteristics, technical diversity, and service complexity of satellite communications have evolved over the last three decades, the regulatory environment has become equally complicated as well. If the world of terrestrial mobile and wireless communications seems difficult to comprehend in terms of standards, regulations, and overseeing agencies, then satellites are clearly even more challenging.

There are officially some 17 satellite services as defined by the International Telecommunication Union (ITU) and listed below.[2] These diverse space communications services, some of which are recognized only in footnoted radio allocations, vary greatly in commercial value and degree of actual use.

(1) Fixed Satellite Service (FSS)
(2) Broadcast Satellite Service (BSS) (also commonly known as Direct Broadcast Satellite Service (DBS) or even small dish television)
(3) Broadcast Satellite Service for Radio (BSSR)
(4) Radio Determination Satellite Service (RDSS)
(5) Radio Navigation Satellite Service (RNSS)
(6) Inter Satellite Service (ISS) (also known as Intersatellite Links (ISL))
(7) Mobile Satellite Service (MSS)
(8) Aeronautical Mobile Satellite Service (AMSS)
(9) Aeronautical Radio Navigation Satellite Service (ARSS)
(10) Maritime Mobile Satellite Service (MMSS)
(11) Maritime Radio Navigation Satellite Service (MRSS)
(12) Land Mobile Satellite Service (LMSS)
(13) Amateur Satellite Service (ASS)
(14) Meteorological Satellite Service
(15) Space Operation Service (SOS)
(16) Space Research Service (SRS)
(17) Earth Exploration Satellite Service (EESS)

In contrast to localized terrestrial mobile communications, satellite services can be offered on a local, national or subnational, regional, or global basis. In the case of global satellite services an operator faces enormous regulatory challenges. Such an operator must have licensing authority, local frequency allocations, international approvals for orbital locations and frequencies, earth station access, and possibly even antenna siting agreements under local zoning authorization concerning the size, appearance, and power of antennas.[3] It is because of this complexity that consortium arrangements with local partners in participating countries, such as INTELSAT and INMARSAT, have proved so popular around the world.

Authority to interconnect with the Public Switched Telecommunications Network (PSTN) may be among the most difficult regulatory hurdles to be surmounted in placing a satellite service into operation. In short, in seeking authority to operate a global satellite system there may be many hundreds of approvals that must be obtained at the local, national, regional, and global level and in some cases it is more than just meeting a particular technical standard or guideline. Barriers to trade in satellite communications may be in evidence in terms of high tariff barriers for imported earth station equipment, complicated testing programs to limit importation of equipment, inability to obtain the authority to operate in needed frequency bands, or even denial of the right to interconnect with the PSTN. In many countries one may not be allowed to own or operate earth station terminals except by undertaking a very rigorous and time-consuming series of tests and regulatory steps. Even then one might find that it is simply not economically feasible to operate a network in some countries. In others in can even be explicitly prohibited to own terminals. Even when clear licensing or other arrangements for earth station ownership are specified, such as under the European Commission Green Paper, the process may still be very difficult and costly.[4] Private international or regional satellite systems such as PanAmSat and Orion have found these regulatory and licensing hurdles particularly daunting to their efforts to create competitive satellite systems.

For many years the regulatory barriers to operating international earth station facilities in the form of a global network were so high that the only practical way was to rely on local PTT or local PSTN operators for earth station ownership and operation in territories where monopolized telecommunication organizations existed. The process of global satellite communications was first established through the INTELSAT global satellite system. IN-TELSAT was established as the global consortium for FSS services in 1964 and was reconstituted as an intergovernmental international organization in 1973. It combines some 130 member countries (and almost 200 countries and territories as users) to own and operate a global network of 20 satellites on a collective basis. These 130-plus owners thus represent these same still largely monopolized telecommunications organizations. The same structure is largely mirrored in the INMARSAT organization for international mobile satellite services although INMARSAT's membership has grown to about 75. Today, however, the situation is much more complex with the evolution of competitive private satellite systems for international and regional traffic. Further competitive practices and access within the INTELSAT and INMARSAT satellite systems are evolving and there is at least discussion of even privatizing INTELSAT and INMARSAT. Today the earth station access procedures are clearly more flexible and accommodating to new entrants with competitive access now allowed in the United States and throughout most of Europe.[5]

Today there are over 200 geosynchronous satellites in orbit providing global, regional, and national coverage that includes a very wide range of services. Some systems are private competitive networks offering global or regional services such as PanAmSat, Orion, or Asia Sat. Others are providing national competitive services, such as the Hughes Galaxy, the GTE Spacenet, or the Morelos system of Mexico. Numerous additional private networks are planned. In some cases national and regional systems such as the Ellipso system in the United States and the ECO-8 system for Brazil and equatorial countries may employ nonconventional networks using new types of orbits. These, however, are more likely to be the exception rather than the rule.

Some systems are national satellite systems that also provide regional services, such as the Indonesia Palapa system, the Australia Aussat or Optus system, or the Hispasat system of Spain. Even the INTELSAT, INMARSAT, and the EUTELSAT satellite systems, which were created in an environment where national monopolies predominated, have now adapted to the new competitive atmosphere and allow competitive access by multiple organizations as national governments so authorize. These systems can and do provide national services as well.[6]

The purpose of this chapter is to provide certain basic fundamentals of satellite communications and then to examine in greater detail the regulatory procedures that apply to satellite communications at the national and international levels.

5.2 THE FUNDAMENTALS

The basic premise of satellite communications is quite simple. The satellite acts as an active relay to transmit messages over large distances. The fundamental components of a satellite transmission are shown in Figure 5.1.

In some cases the signal is simply translated from the up-link frequency, filtered, amplified, and retransmitted back to earth. On the satellites with the latest technology, the signal may be digitally processed and regenerated on-board the satellite to achieve an enhanced signal. At this point, however, several key observations are particularly noteworthy.

To function properly and usefully, all modern satellites must have "approved frequencies" for both the up-link and the down-link. These typically involve national and international approvals by the national governments and the International Telecommunication Union (ITU). There are also standards and rules for the use of these frequencies. These include guidelines for minimizing unwanted off-axis emissions or "side-lobe" transmission that could interfere with other satellite systems or terrestrial microwave systems. In some cases formal intersystem coordination procedures are required both under national and international regulations.[7]

Furthermore, all satellite systems must have their orbits registered and approved at the national and international levels. These orbits are also subject to intersystem coordination procedures. In recent months, however, the Chinese APSTAR 1 satellite was launched into geosynchronous orbit at a location of 131 degrees east in close proximity to the Rimsat (i.e., the joint Russian/U.S. venture) and the Japanese NTT N-Star satellite without prior coordination or registration through the ITU systems. This suggests that international procedures must be greatly strengthened as more and more competitive satellite systems are launched. Fortunately arrangements were made to place the APSTAR in an alternate position and a major international dispute involving China, Russia, the United States, and Japan was avoided.

Satellites come in a great variety of sizes (25 to several thousand kilograms) and their telecommunications capacities vary greatly as well. Some can only store and forward e-mail while others can carry hundreds of broadcast-quality television channels in real time. Some are spin stabilized and look like a cylinder with an antenna system on top, while others are three-axis stabilized by means of momentum or inertial wheels. These, as can be

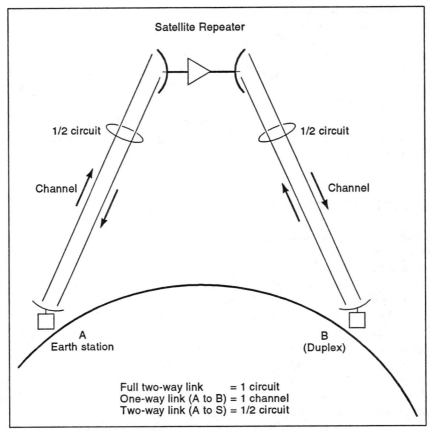

FIGURE 5.1   Basic Components of a Satellite Transmission

seen in Figure 5.2 look more like birds with the solar array system representing the widely spread wings. For scale purposes a human form is shown with the various satellite types.

The performance and throughput capacities of satellites increase with more on-board power and the ability to concentrate beams to smaller and smaller locations by means of higher-gain antennas. Figure 5.3 shows some conventional satellite beam patterns that the regional satellite systems known as EUTELSAT and ARABSAT have used. Today, as can be seen in the illustration provided in Figure 1.8 with regard to the Motorola Iridium satellite systems, much "tighter" beams are now being planned to boost frequency reuse (as in terrestrial cellular systems) and to boost power levels as well.

Today satellites can provide a wide range of services ranging from fax, data, multimedia, and voice, to imaging, television, and HDTV. Satellites can provide very high throughput services via sophisticated very-large-aperture earth stations down to very-small-aperture terminals (VSAT) for lower volume and remote services. In the future, DBS and higher-powered satellite services will be delivered to even smaller ultrasmall-aperture ter-

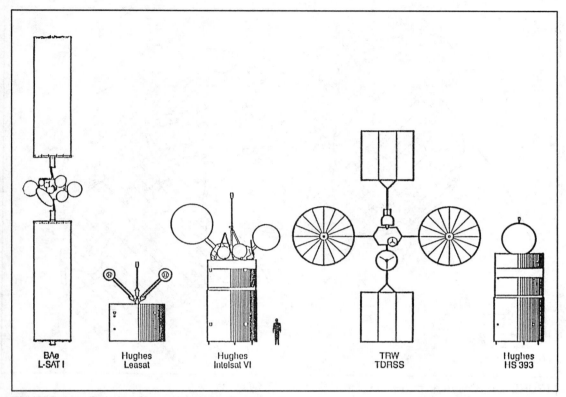

FIGURE 5.2   Relative Sizes of Satellite Spacecraft

minals (USATs). The range of satellite service just to support global entertainment is shown in Figure 5.4.

Finally, earth station antennas are also subject to technical standards and approval procedures. Today receive-only antennas require little coordination while most antennas that transmit and receive are subject to more strict controls. Small antennas that radiate a lot of power are particularly critical because the small reflectors create very broad beams, with the maximum opportunity for sending interfering signals to adjacent satellites. The problem of unwanted side-lobe transmissions, which increase as aperture size decreases, is shown in Figure 5.5.

Another key factor involving earth stations is that the size of the aperture is inversely proportional to the square of the wavelength being used for transmission. This is because the antenna reflector can "capture" a much larger number of the incoming waves as the wavelength becomes smaller and smaller. The shape of the reflector for similar reasons, however, must be more and more perfectly shaped to concentrate the signal to the antenna's feed system. As the frequencies increase another major change also occurs. This is the precipitation attenuation effect, which serves to deflect radio signals from their broadcast pathway. The distortion effect unfortunately becomes exponentially worse as the signal frequency increases. The measurements taken of precipitation effects for the most frequent-

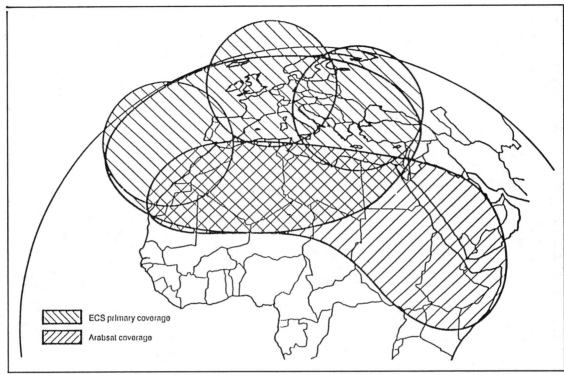

FIGURE 5.3   Coverage of ARABSAT and EUTELSAT Systems

ly used satellite band-widths—namely the C Band (6 GHz/4 GHz) and the Ku Band (14 GHz/12 GHz)—are shown in Figure 5.6.

In general the rate of innovation in satellite communications is moving at a very fast pace. Over the last decade, the following developmental path has been followed to make satellites less of a bent pipe and more of a highly capable and "intelligent" telecommunications platform with an expanding range of functions with higher and higher capacity. The trend line in terms of improved economic performance as well as in key technological innovations is reflected in Figures 5.7 and 5.8, respectively.

Key technical issues that create regulatory concerns with regard to satellite services include such areas as:

(a) Intensive use and reuse of frequencies, particularly at very-high-power levels.

(b) Broadcast systems that operate at particularly high power levels and sometimes over large geographic areas and sometimes with unwanted "spill over" into adjacent countries.

(c) Global beam transmissions that limit the use of frequencies over very large areas.

(d) Polarization techniques for frequency reuse purposes that lack sufficient discriminatory powers to allow the full amount of frequency reuse that would be considered desirable or theoretically achievable.

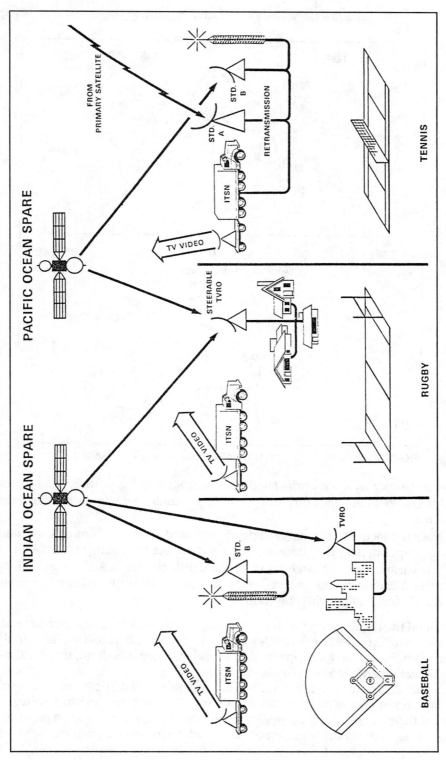

FIGURE 5.4 International Television Sports Network Using INTELSAT Satellites

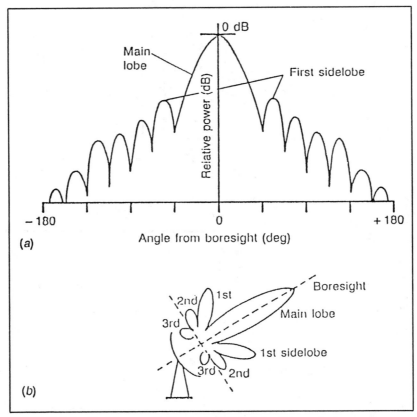

FIGURE 5.5 Problems with Unwanted Side-lobe Transmissions

(e) Interference with terrestrial telecommunications, radar, or radio-telescope services.

(f) Interference between and among space communications and other space-based systems.

(g) National security- and national defense-related issues concerning the unauthorized interception, jamming, or interference with space communications systems.

(h) The various types of satellites in terms of different service offerings, different modulation and encoding systems, and different orbital configurations are often technically incompatible with one another.[8]

In general the regulatory process with regard to satellites is difficult for several key reasons. First, there is the issue of satellites interfering with other satellites, terrestrial communications systems, radio-telescope, and even radar systems. Second, there is the problem that earth station antennas can cause interference to these systems as well. Third, there is the issue that different classes of satellites for fixed, mobile, broadcast, radio determination, data relay, and intersatellite links can and do interfere with one another. Fourth, satellites because of their technical characteristics are essentially often only able to provide an international or regional service and cannot easily avoid radiating signals into other countries.

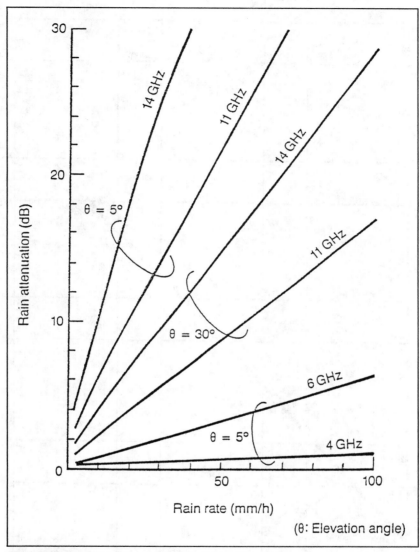

FIGURE 5.6   Rain Attenuation Effects

(This means that while most terrestrial systems of telecommunications are local or national in scope, satellites typically involve many countries, transborder spill over and international frequency allocation procedures.) Fifth, the addition of intersatellite links (ISLs) also allow satellites to bypass national networks, which also leads to regulatory issues and problems.

In short, if one had wanted to create a technology that would give rise to international regulatory issues, there are few inventions that could have done the job better than the communications satellite.

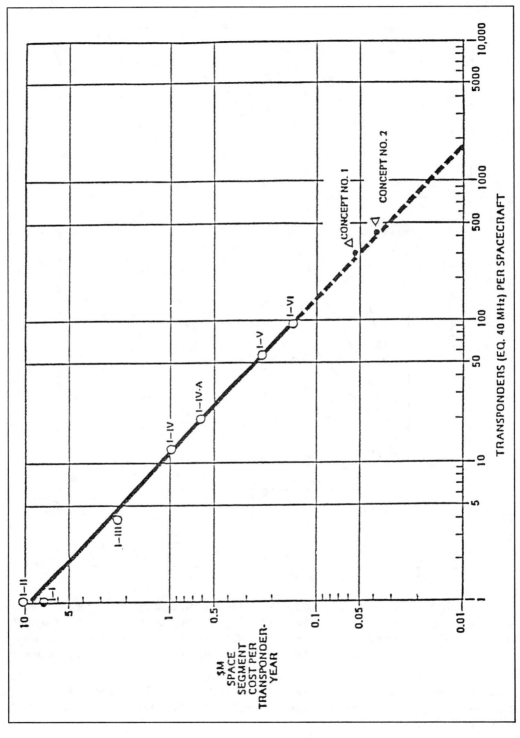

FIGURE 5.7 INTELSAT Series Demonstrates Economy of Scale in Communications Spacecraft (Past and Present)

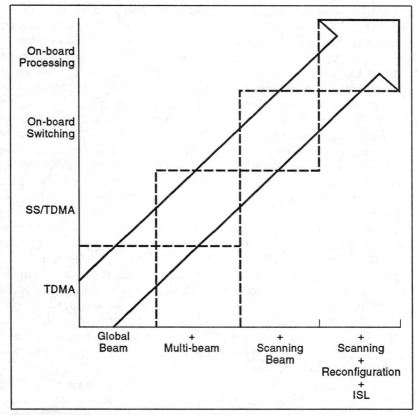

FIGURE 5.8 Process Toward Switchboards-in-the-Sky

5.3 THE REGULATORY ENVIRONMENT FOR SATELLITES

The remainder of this chapter will thus address the regulatory and standards issues concerning satellite communications in the following order: (a) frequency allocations, (b) orbital assignments, (c) intersystem coordination, (d) earth station operation and approvals, (e) licensing of satellite systems, and (f) interconnection of satellite systems with other satellite systems and terrestrial systems.

5.3.1 National and International Procedures for Satellite-Related Frequency Allocations

The process of obtaining frequency allocations for a satellite system essentially involves two aspects. First, there is the international process whereby frequency bands are approved for use for satellite applications on a primary, secondary, or noninterference basis through the International Telecommunication Union. This initial allocation process is actually accomplished through something called the World Radio Conferences (WRC) or in some instances Regional Radio Conferences (RRC). These are, in fact, plenipotentiary conferences of the ITU System that have treaty-making powers. If new frequencies are allocated for new use there may be a waiting period of several years before the reallocation process is completed. Once these bands are approved through the WRC or RRC process, then intended use of these bands must be registered on the Master Frequency Registry by a formal filing by a national administration formally assigned this responsibility. This registration is a long and complicated procedure that takes many months or even years to complete. The ITU officials actually publish each proposed filing and send it to all ITU members, which is nearly 200 countries. They then await any comments about potential interference of the proposed satellite network with regard to any existing or proposed satellite network or terrestrial system. If there are comments indicating interference, then a formal process is followed to resolve these problems. This process of intersystem coordination is described below in some detail.

The national part of the process is equally important. Most countries' agencies that are responsible for being the ITU administration controlling the registration of national satellite networks will not proceed until formal procedures are completed within their own jurisdiction. The case of regional and international satellite systems is somewhat more complicated in that usually a single national administration is designated to act as the registering agent for the international organization. International satellite frequency registrations with the ITU in the case of INTELSAT is through the United States, for INTERSPUTNIK it is through Russia, for EUTELSAT it is through France, and for ARABSAT it is through Saudi Arabia. In these cases the proposed frequencies are filed in the sense of being an official conduit.

This means that the United States might file an INTELSAT document with the ITU as the filing administration, but that the United States might subsequently file objections with the ITU in terms of potential frequency interference with U.S. national satellite systems. In the case of the United States, comments with regard to international frequency registrations are coordinated through the Federal Communications Commission, but the actual transmission of comments from the United States to the ITU is made through the officially designated administration, which is the U.S. State Department.

There are a wide range of comments that might be filed in response to a new international satellite filing. The FCC in the United States is most likely to hear about the potential for interference into an existing or planned satellite system as noted by the owner or operator of those systems that may be affected. This means that satellite operators must constantly review new filings or revised filings that many cause interference. Second, there may well be a problem with the proposed frequency band in that it actually violates the agreed "rules." There are numerous possibilities here. It may be a matter of out-of-band emissions, unwanted characteristics, lack of sufficient polarization discrimination, or most fundamentally, a violation of the agreed frequency use provisions. A frequency band may be approved for one or more of the ITU regions but not all three. Some countries may register a footnote exception for the use of a particular frequency in their country. At one time, such footnotes were rare, but today they are increasingly common. The greatest number of instances of national footnoting is when they are protecting an existing military application for, say, radar services. As but one example, the Russian Gonets store- and forward-data relay satellite system is blocked from service to the United States because of preexisting frequency usage for U.S. military applications.[9]

In some countries, including the United States, the process of getting a national license to operate a satellite system is a long and complicated process. Difficulties in the international coordination processes of the ITU are thus usually detected and hopefully resolved even before the international filing with the ITU is made. Other countries have much more streamlined procedures and filings can often be rather freely and quickly made. This leads to resolution of the problems at the international level. Within the last decade, another key development has occurred. These might be called "flag of convenience" countries for new satellite systems. In one instance, the country of Papua New Guinea agreed to file for a system known as the Pacific Satellite System (PACSTAR) with the ITU in exchange for free use of one transponder on the system for their domestic communications needs. In terms of planning, design, manufacture, and operation, this satellite system would have been under the control of a U.S. corporation, but the nominal national administration for PACSTAR would be Papua New Guinea.[10]

In another more extreme case, the small Pacific island country of Tonga has authorized a corporation known as Tongasat to "market" satellite frequency bands and orbital locations for lease essentially to the highest bidder. A U.S. businessperson with a long history of involvement in the U.S. and INTELSAT satellite industry actively monitors upcoming or pending new filings for satellite systems that might be forthcoming. As filings are made at the national level, generic satellite filings are quickly generated and filed with the ITU through the Kingdom of Tonga. Since national review procedures for new satellite systems often take many months to complete, the Tongasat filing is almost always ahead in terms of priority. There is nothing in the existing procedure that obligates filers to actually proceed with their indicated plans. Furthermore, Tongasat can on speculation choose what seem to be desirable orbital locations, frequency bands, or even global constellations of low or medium earth orbit satellite systems and file for these hypothetical systems as well.[11]

There are two rather different ways to look at this current rather confusing international regulatory system for registration and coordination of satellites. One way is to find fault with the lack of accountability and commitment to implement systems that are filed with the ITU. (The current system places the responsibility on national administrations to file systems that are actually intended to be deployed.) The other way of viewing the situation is that the new emergence of brokers, facilitators, or agents who facilitate rapid access to desired satellite frequency bands and orbital locations can actually speed up access to the needed resources and as long as the "brokerage fees" are reasonable there is ultimately little harm done. Most established satellite system owners and operators, however, are extremely upset with these "flag of convenience" entrepreneurs who have exploited a lax set of international procedures for financial gain.

Today the existing procedures are already being revised to place less emphasis on the concept of "first come, first served" that has governed the satellite arena for some 30 years. There will likely be other changes that may involve substantial filing fees, penalties for instances of significant noncompliance with filings, or other barriers to the filing of "paper satellite systems" that are not backed by serious intent.[12]

The national procedures concerning the establishment of new satellite systems vary widely from country to country. In many instances the government itself plans, designs, and seeks permission for national satellite systems. This makes the process quite straightforward and there is no issue of competitive applications. In the United States, however, the situation is far more complex. First of all there may be governmental or military satellite systems that are coordinated within the official governmental processes. There may be special combined governmental and civilian projects that require special coordination procedures. Finally, there may be a need for special comparative evaluation where there are many different corporate filings. These procedures, which are typically defined by the FCC, can vary from one circumstance to another. In one case of applications for national geosynchronous mobile satellite services, filings were made by eight different commercial organizations to design and deploy such a system. After over a year of comparative hearings the FCC decided to request six of the applicants to form a consortium to proceed with a single cooperative project. Two of excluded applicants sued in the federal courts and were successful in being included in the consortium. In other more recent cases the FCC has tried to maintain competitive applications, but has tried to resolve key issues of how to best distribute the limited available frequencies for low and medium earth orbit satellites for mobile services. An attempt to arrive at a negotiated settlement between the numerous applicants was unsuccessful. In this instance, a NASA official appointed by the FCC presided over the conflicting petitions, accounts of which follow.

On one hand the Motorola Iridium Satellite, the first to file for such a system, maintained it needed to use well-proven TDMA technology and bidirectional service to use the limited available frequencies, in effect, twice. The other so-called big LEO satellite systems, namely Globalstar, Odyssey, Aries, Ellipso, and so on, petitioned to use the same band with higher efficiency CDMA technology and unidirectional service. After a year of negotiations among all concerned parties the formal result was that Motorola proposed segmenting the band with half for Motorola's Iridium System, which would use TDMA in a bidirectional mode, and the other half for everyone else, using CDMA. The remaining petitioners, i.e., everyone but Motorola, proposed that all systems operate with CDMA and that the bands be shared among all applicants.[13] In this case the negotiated rule-making NRM process essentially failed. Today the status of frequency allocations for the so-called "big LEO" systems is still in some disarray. Although the FCC has opted for a compromise approach of splitting the band between TDMA and CDMA, the European Commission is still unhappy with the United States unilaterally deciding on how frequencies for low-orbit mobile systems will be allocated. This could well lead to problems at the 1995 World Administrative Radio Conference.

The FCC is still inclined to pursue the Negotiated Rule Making approach, however, in that it is now faced with the problem of the terrestrial and space-based sharing of the 28-GHz band. At this time there are a number of experimental trials for the terrestrial applications, such as US WEST's San Diego trial and a number of space-based users and applications such as the NASA ACTS experimental satellite and the pending proposals of the Teledesic Mega-LEO system, the Hughes' Spaceway System, and the Norris Norstar System. In this case since many of the applicants are not actually in head-to-head competition with one another and because there may be some viable technical solutions vis à vis space and terrestrial applications, the prospects of a successful "negotiation" may be better than in the case of the "big LEO" system's NRM.[14]

What was particularly disappointing to Motorola Iridium was their unsuccessful attempt to achieve a status known as a "pioneer preference," which would have given them their frequency allocation and a "regulated head start" over their competition. This is a procedure adopted by the FCC in the 1980s whereby an applicant with a totally new product or service and with totally unique technology seeks a special status, which has two key components. One is the award of needed frequency bands for the service and the other is a six-month head start. When Motorola filed in the mid-1980s the thought that they might be awarded pioneer preference status seemed plausible. Certainly at the time the idea of a global constellation of low earth orbiting satellites with intersatellite links and hand-held transceivers was a highly innovative concept. A number of well-established satellite engineers were even skeptical that it could be done.[15]

In a very short order of time, however, a number of competitive applications for similar types of satellite systems were filed. Experts recalled that the ITT corporation, COMSAT, and others had done studies for the INTELSAT system some 20 years earlier for similar types of low-orbit systems. The final result was that the FCC, under a barrage of testimony and competitive applications that showed comparable capabilities, decided not to award a preference to Motorola.[16]

To file successfully for a satellite system to the FCC requires a major and rather painstaking effort. The application must spell out in some detail the satellite system's space segment and ground segment characteristics in some detail. This includes frequency bands, frequency reuse concepts, power levels, launch arrangements, communications subsystem including modulation and encoding techniques, intersystem interference and intermodulation calculations, tracking, telemetry, and command subsystems, plus a detailed description of all ground-based facilities, such as master control facilities, ground-based intersatellite links, gateway stations, user transceivers, and interface arrangements for interconnection with the public switched telecommunications network (PSTN). This is, in fact, only the beginning. There is also a need to present detailed financial and business arrangements, market assessments, management, and implementation schedules along with a statement of technical credentials. In competitive applications the filing often seeks to state why a particular system is more technically, operationally, and financially worthy of selection than other applications. Often a case is made as to why a system should be selected because it uses the available spectra more efficiently, provides greater capacity, allows operation with more compact or cost efficient ground transceivers, or will provide a service that costs less than the competition. Since these procedures are handled in a semijudicial manner, other systems have the right to make counterstatements and rebuttals. Thus in the U.S. system the filings are usually written by attorneys rather than engineers and adversarial debate can assume as much importance as the technical design.

This system has been frequently criticized for being overly legalistic, time consuming, and even unfair in submerging technical merit under lawyerly debate and hyperbole. The competitive process does, however, seem to give rise to innovative ideas, entrepreneurial initiative, and in many cases the "survival of the fittest." Many defenders of the system would note that it allows for entry by new start-up ventures and places a premium on new ideas and totally different systems concepts. The area where there seems to be some agreement among critics and defenders of the current system is that the process is tremendously time consuming and perhaps overly complicated. Certainly if the current procedures could be speeded up, it would seem highly desirable. This would allow U.S. filings to be sent forward into the ITU process much more rapidly rather than always being toward the end of the list of international applicants.

5.3.2 National and International Procedures for Orbital Registration

The processes of obtaining frequency allocations and registration of orbital characteristics are in fact closely intertwined. In attempting to simplify these processes, the two steps have been separated in this discussion. The two aspects that best describe a specific satellite system are the frequency bands utilized and the orbital characteristics of its satellite networks. With these parameters it is possible to track and monitor any communications satellite. A satellite for communications purposes uses a variety of different orbits. The most popular one to date is the geosynchronous orbit, which allows continuous communications between a satellite and earth station antennas pointed to the satellite's fixed location above the earth. Today there are over 200 communications satellites in geosynchronous orbit. These typically use the L, S, C, X, and Ku Band frequencies. As noted above there is also increasing interest in the use of the Ka Band with multiple filings for the use of this band as well. The band that is most congested is that of the C Band, where satellites virtually circle the globe with very little spacing between these networks in the 6 GHz/4 GHz band. This congestion of the geosynchronous orbital plane is really not an issue of physical propinquity but rather the frequency interference that occurs at relatively close range.[17]

One degree of orbital separation in geosynchronous orbit is, in fact, equivalent to a distance of some 735 kilometers or 460 miles between the adjacent satellites. There is thus seldom an issue of orbital collision so much as radio signal interference.

There are many other orbits of interest to satellite communications. A global constellation of satellites properly deployed can provide full global coverage. The lower the orbit, the more satellites are required to accomplish this purpose. The medium earth orbit configuration (i.e., circular orbits of 8000 to 13,000 kilometers or 5000 to 8000 miles) such as that proposed by TRW's Odyssey system will provide continuous global coverage with a network of 12–18 satellites. It is likely that the INMARSAT Project 21 will also use a similar global constellation. At low earth orbit (i.e., 1500 kilometers and below) the number of satellites required to provide continuous voice services increases rather dramatically. The number of satellites required for low earth orbit voice service is typically in the 50–70 range, even though concepts for so-called mega-LEO systems have also been envisioned. One such system, known as the Teledesic Communications, has proposed as many as 840 satellites. This system envisions some 40 satellites in each plane with 21 different planes covering the earth's surface. In this case, the very large number of satellites was not derived on the basis of achieving total earth coverage, but rather on the basis of achieving large economies of scale and deploying huge telecommunications capacities using much higher frequency bands, namely the 30 GHz/20 GHz Ka Band. The Teledesic system design, in theory, would have 100 times the capacity of say the Iridium Satellite System. Financing for the Teledesic System has not yet been secured, but the active participation of Craig McCaw and William Gates, two of the wealthiest men in America, is an important indicator of the potential financial viability of the project.

Although the geosynchronous systems plus the new low and medium earth orbit circular global constellations represent the vast majority of existing and proposed satellite systems in orbit, or planned for the next decade, there remain many other options. These include the elliptical orbit or highly elliptical orbits, which are more typically used for earth remote sensing, and the nongeosynchronous equatorial orbits, such as the proposed Brazilian ECO-8 Satellite System, as well. Then there are special variations on these themes, such as the so-called "loopus" orbit, the specialized elliptical orbits of the Ellipso system designed to create coverage focused on the U.S. mainland, and super-synchronous orbits for special military applications.

In terms of broad trends, well over 95 percent of the communications satellites now in orbit and those planned for the near- to medium-term future will likely follow the general guidelines outlined below.[18]

(a) **Fixed Satellite Service (FSS) systems** for domestic, regional, and international coverage will be deployed in the geosynchronous orbit at least over the next decade. Some may be deployed in equatorial circular orbit, however, such as the Brazilian ECO-8 satellite system.

(b) **Broadcast Satellite Service (BSS) systems**, or DBS satellites, will generally be deployed in geosynchronous orbit at least for the next ten years. Again, it would be possible to use the equatorial circular orbit.

(c) **Mobile Satellite Service (MSS) systems** will be deployed in several different types of orbits. These will include geosynchronous, low and medium earth orbit, and elliptical orbits. (Of the low and medium earth orbit satellites designed for mobile services there will be two types. One type, sometimes known as the "little LEOs" will operate at lower frequency bands and offer a lower band-width store- and forward-data or e-mail type service. The other type, sometimes known as the "big LEOs" will operate at higher frequency bands (i.e., around 2 GHz) and will offer a continuous or voice-based service. Little LEOs will generally be smaller, less powerful and require much fewer satellites (i.e., about one-third the number of the big LEOs) to achieve global coverage to support messaging services).

(d) **Other types of satellite systems**, which include (i) military- and defense-related satellites; these can be in low, medium, elliptical, geosynchronous, and even super synchronous orbits; (ii) radio determination or navigational satellites, which would most likely be in medium earth orbit although navigational receivers are often also placed on low earth orbit satellites; (iii) data relay satellites, which are usually placed in geosynchronous orbit; (iv) meteorological and remote sensing satellites, which are often in medium, polar, and sun synchronous orbit; and (v) intersatellite link systems, which may involve geosynchronous, medium and low earth orbit systems.

These rules are not inviolate, but they do characterize the current situation and will likely represent the future situation at least a decade into the future. The clear pattern is that satellites for many different applications are gradually "saturating" the available orbits as the demand for space services continues to grow even with the great popularity of fiber optic cable systems. This process of "saturation" does not mean filling all of the physical space within an orbit, but rather, a saturation of available frequency bands.

Today, the C Band (i.e., 6 GHz/4 GHz) in geosynchronous orbit is essentially already filled. Expansion of satellite services in this band must be essentially by means of frequency reuse and by digital compression techniques. The Ku Band in geosynchronous orbit (i.e., 14 GHz/12 GHz) is only slightly less congested. Expanded use of this band is thus largely dependent upon frequency reuse and digital compression techniques as well. The allocations for so-called little LEO and big LEO mobile communications satellite systems at 2 GHz and below are all very limited in band-width. Essentially the currently proposed systems from the United States, Russia, and Europe, plus the anticipated filings for new international systems such as for INMARSAT and Japan will more than claim all of the available spectrum even with intensive frequency reuse techniques.

There are, in fact, only a few small bands below 20 GHz that might have some growth potential and even these are limited. There is an "S" band allocation at 2.5 GHz/2.4 GHz for satellite community antenna educational distribution. Further, there are some military communications satellite bands in the "X" band at 8 GHz/7 GHz, which is not fully exploited. Additionally some DBS feeder and distribution links around 18 GHz have yet to be fully used. Despite these few possibilities and some other odds and ends, the major new frequency band with major service expansion possibilities is clearly the 30 GHz/20 GHz band, which may be eventually used for FSS, BSS, and MSS services. Experimental satellite systems such as the United States' ACTS program, the Italian ITALSAT project, and the Japanese ETS VI experimental project have attempted to determine how to overcome the problem of rain attenuation at these very high near millimeter wave frequencies.

Commercial systems, based upon these experimental programs have now begun to commit to such high frequencies and as noted earlier there are now at least three U.S. systems that have been filed with the FCC. These are the mega-LEO system of Teledesic, the Norris Satellite System (Norstar), and the Hughes Space Way System. All three have indicated plans to deploy operational 30 GHz/20 GHz satellite systems in the twenty-first century. Finally there is consideration of a new broad band allocation in the millimeter wave band that might be as large as a 5 GHz band, perhaps in the 40–50 GHz region.

The four key trends in satellite communications are toward the use of different types of orbits, toward the use of higher and higher frequencies, toward increased switching and intelligence on-board the satellites, and toward more intensive patterns of frequency reuse, particularly within those bands that are already close to saturation.

The ITU will be challenged to cope with this exploding technology and the surging demand for orbital assignments. More and more sophisticated computer simulation systems will be needed to assess the interference between and among adjacent satellites in geosynchronous orbits but also between and among low, medium, and elliptical orbits as well. This will also mean that formal intersystem coordination procedures will become more complex and difficult as well. In the past, intersystem coordination procedures often involved only the two closest satellite networks. In the future dozens of satellite networks may be involved not only in earth-to-space and space-to-earth links, but also involved in space-to-space link coordination as well. At the Region 2 Administrative Radio Conference on Broadcasting Satellites held in the 1980s, the first computerized systems model for assessing channel capacity and adjacent satellite interference employed to determine BSS channel allocations and orbital positions was successfully used. Since that time, increasingly sophisticated computer modeling has been used for this purpose.

5.3.3 Intersystem Coordination Procedures

The process of intersystem coordination is governed by several sets of rules. First, there are the rules and regulations of the ITU as agreed upon at the World Administrative Radio Conferences, then there are the rules and stipulations of the various international satellite treaty organizations, such as INTELSAT, INMARSAT, EUTELSAT, and ARABSAT. These rules are binding for the member countries of these organizations and to a lesser degree for the nonmember users of these organizations. Particularly, in the case of INTELSAT, which has over 130 member countries, the intersystem coordination procedures are almost like an international convention. INMARSAT, with over 70 members, also has a wide range of influence as well. Since most of the procedures on international satellite system coordination derive from the INTELSAT Agreement, and especially Article XIV of that international agreement, it is perhaps most useful to review these provisions.[19]

The first element to note is that the provisions of Article XIV of the INTELSAT Agreement have become quite controversial, and, in fact, have been amended in practice in recent years to reflect the transition from monopoly oriented telecommunications to competitive systems. The first step was a decision by the INTELSAT Assembly of Parties to delegate to the INTELSAT Board of Governors to take all actions related to intersystem coordination wherein the international traffic levels were below certain minimum levels. This allowed the Board of Governors to complete the technical and economic coordination of all systems except for major international systems planning to carry a large amount of public switched telecommunications traffic.

Further pressure coming from the United States and Europe has now led to a practice whereby INTELSAT intersystem coordination procedures focus on the technical issues of satellite carriers that could create serious interference into adjacent satellites and have deemphasized the most controversial Article XIV (d) provisions that test "economic compatibility" of new international systems. In effect, INTELSAT has agreed that it will not seek to constrain new competition in international satellite services in exchange for the tacit understanding that governments will not seek "excessive" new entry into this field. This new "status quo" in the field of international satellite system intercoordination is important in that the other satellite systems in the form of EUTELSAT and INMARSAT are no longer seeking to enforce their economic coordination procedures either.

Certainly, no one would deny that technical coordination of satellite systems is crucial. The resolution of potential interference problems is clearly best achieved when the satellite system is being planned and designed rather than after a satellite is manufactured or launched. Since rather precise equations and even sophisticated computer simulations can now predict the characteristics and levels of interference of satellite systems and their earth stations with a high degree of accuracy, the process of coordination is much more clear-cut and effective. The variables that constitute the major elements of the intersystem coordination process are also clearly definable. They include:

(a) The satellite's respective orbital locations and relative spacing from one another.
(b) The number, frequency range, and power levels of the transponders as well as their linearity, and any special capabilities like reconfigurability or retuning into different frequency ranges.

(c) The type of modulation, encoding, and carrier sizes that will be used.

(d) The system of frequency reuse and polarization that will be used.

(e) The precise characteristics of the antenna systems and their side-lobe transmission performance.

(f) The precise beam sizes and pointing characteristics of all antennas.

(g) The performance characteristics of all earth segment facilities operating to the satellite network.

(h) Any special characteristics such as intersatellite links, hybrid systems for cross strapping between frequencies, special high gain settings, and so on.

In theory, the process of intersystem coordination is one of first understanding how transmissions from a satellite and its operating earth stations can create unwanted interference into other satellites or its earth stations, which may be operating in the same frequency band or harmonics thereof. The next step is then to minimize or eliminate that interference by a variety of techniques. This might be lowering power levels, shifting frequencies from say the interfering beam to another beam; altering modulation, encoding, and multiplexing methods; inserting filters; changing the performance characteristics of earth stations; using different types of polarization techniques; repositioning the satellite; or creating different beam patterns. This seemingly rather technical and objective exercise has, however, some other components. There are economic implications in the technical coordination process in terms of design changes, altered operational practices, or reduced performance.[20]

The change of a satellite system design to accommodate coordination procedures may not be inexpensive. The somewhat classic case of the Indian domestic satellite system INSAT versus the INTELSAT Indian Ocean satellite over a decade ago eventually required the INSAT system to be relocated to a less favorable geosynchronous orbital location than they had sought. This in turn required that the earth stations for the Indian system have higher gain or performance since there was a lower look angle to their satellite. The net cost of the ground network ultimately cost several millions of dollars more than would have been the case with the most desirable look angle. (In time, India decided to lease satellite capacity from INTELSAT for domestic satellite service to supplement their INSAT network and thus they were able to derive some benefit from the INTELSAT system, but they were nevertheless still unmollified by the economic penalties that they suffered under the international procedures that the INTELSAT Agreement and the ITU procedures imposed upon them.)

The situation can be even more difficult when the parties to an intersystem coordination are competitors with one another. The coordination processes between EUTELSAT and the ASTRA European regional satellite system, between INTELSAT and Orion and PanAmSat, or between Arab and Israeli satellites have been highly contentious. It has been difficult to interpret in terms of where objective technical problems began and where commercial or policy considerations ended. Especially where head-to-head competition is involved one can reasonably expect both parties to seek to have their satellites enjoy the most favorable conditions. This would mean trying to be as close as possible to the target market area in terms of direct look angle. It also means seeking the highest power levels, having the smallest and lowest-cost antennas, using the maximum available frequency bands, and having the lowest-cost satellite system design. These same factors can and do contribute to adjacent satellite interference.

The spread of competition in the field of satellites has not surprisingly led to repercussions in the satellite regulatory framework. There has been increasing pressure to dismantle or dilute the intersystem coordination procedures of the major satellite systems such as INTELSAT, INMARSAT, EUTELSAT, and ARABSAT and to have national governments or the ITU assume these responsibilities. Clearly, the initial international coordination procedures developed for the INTELSAT system under the provisions of Article XIV of their Agreement were designed for a world that has now changed to a much more competitive environment. Furthermore these procedures were then "cloned" into the other global and regional satellite systems. These procedures, which made a good deal of sense several decades ago, were created in a world keyed to telecommunications monopolies and unified governmental regulatory control. Today it is argued that a different process is needed. Actually these concerns have at least two dimensions. Commercial private satellite systems want a "level playing field" that does not have procedures that favor one party over the other. In addition, representatives of developing countries who see the preponderance of satellite technology and systems under the ownership and control of developed countries want to eliminate or lower the barriers or liabilities that "newcomer" satellite systems seem to experience. The previous case of India is a clear case in point.[21]

In the United States and within the European Commission new guidelines for licensing, establishing, and coordinating satellite systems have been adopted to ensure that competitive fairness is achieved. At the international level, the ITU has also begun to modify its rules for registration, coordination, and prioritizing of satellite systems that are registered with the ITU. In particular, "newcomer" rights-of-access and in intersystem coordination procedures with other satellite systems are co-equal with preexisting networks. The onus of accommodation of new satellite systems is placed upon all networks. The balance between according equity to serious new satellite systems, especially those that are the initiatives of developing countries, and to legitimizing "paper filings" for hypothetical or tactical satellite systems that may never actually be deployed is a difficult one to achieve.[22]

In any event, all satellite systems must sincerely seek to achieve accommodation with all other systems. National and international procedures are currently aimed at achieving this result. There are many new tools available that will help. These include new modulation, encoding, and multiplexing techniques. It can also include flexible antenna beam forming, on-board processing and signal regeneration, signal polarization, cellular and spot beam frequency reuse, and side-lobe reduction techniques. On the other hand, other issues such as new orbital configurations, the increase in the total number of satellites in service, intersatellite links, low-gain ground antennas (including hand-held transceivers), and higher-power systems on-board the spacecraft have all served to make intersystem coordination more difficult to negotiate.[23]

As the number of systems and satellite services continues to increase it is reasonable to expect that the stringency and difficulty of intersystem coordination procedures will continue to increase. In particular, standards for out-of-band emissions, precise beam pointing, polarization purity, precise beam contours, cellular-based frequency reuse, and reduction in unwanted side-lobe characteristics may be made even more demanding. Perhaps most difficult to predict will be the standards for satellite orbital separation that will apply in the future. Given the fact that satellite capacity has expanded many thousands of times over the last three decades, it is safe to assume that new and significantly different intersystem coordination procedures will continue to evolve in the twenty-first century. Particularly, new concepts such as the Teledesic mega-LEO system and the planning for High Altitude Long Endurance (HALE) platforms suggest that a number of major changes are still ahead.

5.3.4 Earth Station Approval Procedures

The advent of satellite communications came as somewhat of an unexpected surprise to the international regulatory community. The ITU had had a great deal of experience with regulating radio communications throughout most of the twentieth century, dating from the Marconi experiments. Satellites were clearly a different type of media, but essentially the satellites and their ground antennas were viewed as radio sets. This translated into a process that controlled all satellites and earth stations at both the national and international levels rather strictly as mobile radio transmitters.[24]

This meant that all satellites were subject to national and ITU approvals and registrations and earth stations were essentially restricted to be owned and operated by the national telecommunications monopoly or military agencies. The only exception to this ownership rule was in the United States, where special circumstances were created by the passage of the Communications Satellite Act of 1962. This law, which created a new corporation known as COMSAT, was authorized to become the owner and provider of commercial communications satellite services on behalf of the United States. This entity was structured to have 50 percent of its stock owned by private individuals through the New York Stock Exchange and 50 percent by the international communications carriers, which were at the time AT&T, ITT, Western Union International, and RCA. What the Communications Act of 1962 did not specify was the earth station ownership arrangements. After a formal rule-making process the FCC ruled that there should be an Earth Station Owners Consortium of which COMSAT would be the Manager, but the stations would be 50 percent owned by the international communications carriers and 50 percent owned by COMSAT. Worldwide, the advent of communications satellites meant very tight ownership and control of all earth stations.

In time, however, the situation began to change. As the satellites grew larger in capacity and operated at higher power levels during the 1970s and 1980s, the cost, complexity, and size of the earth station antennas began to decrease. The advent of domestic satellite systems in Russia, Canada, and the United States opened up the concept of large television distribution networks that utilized quite-small-aperture receive-only antennas, known as TVROs or Television Receive Only terminals. Furthermore, plans for mobile satellite systems and direct broadcast satellite systems began to envision even smaller and more compact transceivers or terminals costing only a few hundred dollars for personal use or home use, almost like transistor radios. Clearly, the technology was changing quickly and the regulatory framework seemed unnecessarily stringent and out-of-date. The original idea that there would be only a few multimillion dollar earth stations per country that would be typically owned by the military establishment or the governmentally sanctioned telecommunications monopoly began to seem passé by the end of the 1970s and the start of the 1980s. These developments in many ways mirrored the evolution in the computer industry from mainframes to personal computers.

New ideas mushroomed with regard to how satellite communications and the services, if offered, could be delivered more directly to the consumer. By the start of the 1980s these ideas included: TVRO networks for cable television systems; backyard home television reception; competitive national satellites systems; customer premise antennas for business services; teleports; mobile satellite systems for maritime, aeronautical, and even land-based services; and then, competitive private international satellite systems. It became increasingly clear to the FCC that new earth station approval procedures were needed.[25]

In this respect, the United States' FCC regulatory processes for earth station approvals led the world. This was because the U.S. satellite market, particularly the use of small antennas and TVROs, defined global trends. The FCC actions in 1979 actually led the way and in many ways set the precedents that were subsequently employed in Europe, Japan, and other parts of the world. The FCC thus took a very significant first step in 1979 when it decided that receive-only earth stations were not bound by the mandatory radio licensing requirements of Title III of the Communications Act of 1934. (74 FCC 2D 205 (1979) "Regulation of Domestic Receive-Only Satellite Earth Stations".)[26]

In effect, domestic owners and operators of receive-only terminals were given two options by the FCC. They could chose to proceed on an "unlicensed basis" but in doing so they would be forfeiting any right to objections to interference from any licensed station. The second option would be to seek a license for the receive-only terminal under new streamlined procedures. Under these new relaxed application rules, the need for a construction permit under section 319 (d) of the Communications Act of 1934 is waived as is the need to submit supporting legal and financial qualifications.[27]

The FCC, however, did retain tight control over two-way space antenna systems. This meant that owners and operators of transmit/receive antennas must obtain a construction permit and an operating license. This first involves the filing for a construction permit. This requires a detailed description of the earth station(s) including frequency bands, site availability, channel capacity, cost, facilities and equipment, and so on. It also entails a description of the services to be provided; the technical, legal, and financial qualifications of the applicant; showing of public interest; and necessary waivers under the U.S. National Environmental Policy Act. It also involves a formal earth station coordination process with regard to interference with respect to all registered satellites. Further, a 30-day waiting period must be observed for any comments with regard to environmental waivers.[28]

Assuming that all of these steps are successfully completed and a construction permit is granted, the next step is an operating license. Prior to final completion of the antenna, an application for operating authority is to be filed to certify that all conditions of the construction permit have been fulfilled. The technical requirements established by the FCC with regard to two-way antennas have, in fact, become more stringent primarily in order to accommodate two-degree spacing of satellites and to lessen the effects of off-axis side-lobe emissions involving the use of smaller aperture antennas. There are clearly strict controls still being imposed by these FCC procedures on two-way antennas, but there has nevertheless been a major shift accomplished in that the ownership and control of these facilities within the United States has been greatly extended. Instead of ownership being restricted to the military, COMSAT and a few international carriers, there are now literally hundreds of different owners and operators of two-way antennas and some 4 million owners of

TVROs in the form of home receivers installed to obtain cable television programming in rural and remote areas. Most of these are unlicensed.[29]

In making its decisions on earth station ownership and operation, the FCC has stated that its objective has been to "create a flexible ground environment that would permit a variety of earth station ownership patterns and afford diversified access to space segments." In fact, the owners and operators of interactive satellite antennas in the United States today includes a staggering range of entities including broadcasters, cable television systems, retailers, manufacturers, financial and banking industries, educational and health organizations, and foundations to assume this role. As a new era of personal communications satellite systems emerges and as the Globalstar, Iridium, and INMARSAT become available, it is likely that further relaxation will occur. Relaxed approvals for two-way hand-held communicators will create the next tier of liberalization in the control of space communications.[30]

Although the regulatory liberalization of 1979 was most significant, it is important to note that the FCC has also made other adjustments. These have included authorization of the open resale of satellite capacity, flexible cost sharing and joint ownership arrangements, approval of smaller-aperture antennas, liberalized ownership and operation of international earth stations at customer premises, expansion of the number of authorized international satellite carriers, and greater flexibility to use and access international satellite systems. These international arrangements are now largely made on a case-by-case basis in the case of private network use. The opportunity to apply these streamlined procedures to international networks accessing U.S. public switched network services may be authorized in the not-so-distant future.[31]

These liberalization policies have, in fact, today begun to spread worldwide. The key European Commission Green Paper seeks to establish a comprehensive and consistent set of policies to promote competition and expansion of satellite services throughout Europe. These policies especially have served to promote VSAT services and have in many ways followed U.S. innovations. But these practices are now also being increasingly implemented in national regulatory actions, such as in Germany.[32]

Finally, it should be noted that while international and especially national regulations tend to dominate the control of earth station ownership, licensing, and use, the local municipal governmental authorities can also be of consequence. There are in many jurisdictions zoning, health ordinances, or even franchising agreements that regulate the where, when, and how of earth stations, TVROs, or even DBS terminal installation and operation. These regulations often cover the following: (a) the size, appearance, or even color of an antenna; (b) the siting, elevation, and positioning of an antenna; (c) flux density and power levels; (d) the interconnection with PSTN or even private telecommunications networks; and (e) the resale of telecommunications or broadcast services to others.

5.3.5 Licensing of Satellite Systems

The requirements of establishing a two-way antenna for satellite services is, even in the relatively deregulated U.S. market conditions, still rather a major exercise. The process of obtaining a U.S. license for a satellite system, however, is an exercise of much greater proportion. It is certainly a much more lengthy undertaking. Despite efforts to streamline the process of filing for construction permits and operating licenses for satellite systems, the domestic authorizations can take years to complete. The satellite filing must then go forward into the ITU registration process. As noted earlier, this has led to certain countries serving as "nations of convenience" who can on an accelerated basis serve as a convenient conduit to send an official national filing for a new satellite system forward. Alternatively, this process can be used in a preemptive competitive mode. In this case the action is to use a "country of convenience" to file a competitive "paper satellite" document to "neutralize" the actual satellite system being proposed in another country.

The U.S. procedures for proposing and establishing a new satellite system are outlined below as an illustration of the approval process at the national level. Although some national procedures are more demanding and others are less so, American filing requirements are at least indicative of the requirements of many national governments. In the U.S. process, there are certain threshold issues that dictate much of that which follows. These key issues are whether the proposed system is national or international in scope and whether the satellite's capacity is to be offered as a common carrier or a private network. Also highly important is the type of satellite service that will be offered under the ITU definition of the offering and the associated frequency allocations.

The procedures for private, noncommon-carrier offerings of fixed satellite systems are the most streamlined, but several steps have been taken to advance the FCC procedures in order that international filings with the ITU can be made. In particular, the FCC now allows filings for satellite systems pursuant to Title III of the Communications Act of 1934, as amended, to be carried out in a single one-step process. This means that an applicant can seek in a single step authority for the following: (a) construction authority (Section 319); (b) radio licensing authority (Section 301); (c) launch authority; and (d) if it is a common-carrier service, approvals that would ordinarily be required under Section 214 for new capital investment.

This still requires a very massive filing that covers all of the technical, legal, and financial information outlined above, as well as the information needed for earth station construction and licensing and significant detailed information as well. This application must justify why the public interest would be served by the assignment of the requested frequencies and orbital locations. The precise details required for such an application are specified in Rule 25 of the FCC Rules and Procedures and the FCC's Processing Order. If the filing is not complete in all aspects, the document will be returned to the applicant as unacceptable for filing. Since a completed filing for a satellite system may contain over 100 sections and be 500 to 2000 pages in length, it is normal procedure to retain an experienced law firm familiar with the FCC procedures to prepare and submit the final application.[33]

Special circumstances may apply to certain applications. If a particular applicant feels that their submissions has some special and unique feature based upon the technology or application proposed they can seek a so-called "pioneer preference." This preference, if granted has two advantages. It means the application is accepted by the FCC and a six-month head start in terms of licensing and seeking of international ITU registration will also be granted. This preferential status, however, is not easily obtained. As noted earlier three "pioneer preferences" were granted recently by the FCC with regard to the new terrestrial personal communications services, although payments for the licenses were ultimately required. The application on behalf of the Motorola Iridium low earth orbit mobile satellite system was denied and the Norstar satellite system request to operate in the 30 GHz/20 GHz frequency band is still pending after many months. In short, "pioneer preference" status can yield considerable advantages, but can also serve to bog down the review process and, in fact, allow competitors to catch up during the period of delay. Further details of this process are described in Chapter 7.

A special constraint also exists with regard to filing for mobile satellite systems in low and medium earth orbit. This is largely because the spectrum allocated to these services in the below 2-GHz range is so very minimal, that there are not nearly enough frequencies available to accommodate the proposed satellite systems or their proposed service capacities. In light of these constraints, the FCC felt compelled to utilize the lengthy comparative hearing process. Furthermore, as noted earlier, when this process did not seem to be proceeding well, a further step to seek a negotiated rule-making among the interested parties was pursued to attempt to resolve the issue of TDMA versus CDMA multiplexing systems. This process, which took one year to complete, also deadlocked. Thus despite FCC attempts to streamline the approval process for new satellite systems, the low and medium earth orbit mobile satellite systems have been caught up in U.S. domestic procedures for over three years. In contrast, the French filing on behalf of the Globalstar system was forwarded to the ITU and "country of convenience" filings may go forward to Geneva at any time. Clearly, further work on streamlining regulatory procedures for FCC approval and filing of new satellite systems is still required.

In many countries there is often only a single satellite system that is under consideration for deployment at a time. Further, this may well be a government backed, subsidized, or at least "encouraged" project. In the United States, however, there are an ongoing number of "competitive" domestic and international systems of different architectures, frequency bands, modulation and encoding techniques, and service characteristics to consider. There is often such a sufficient quantity of competitive proposals that the FCC has felt compelled to consider them as a group and on their relative merits by encouraging all interested parties to submit proposals by a set and advertised deadline, usually with a 60-day cutoff. Although this practice has been criticized and even formally challenged in the courts, it has been judicially sustained and is still the most normal practice.

Even if an applicant should complete the application, respond to any supplementary requests for information, go through the FCC review process, and receive a construction permit and a license for a satellite system, the regulatory oversight still continues. There are FCC regulatory requirements that follow after acceptance is given. There are requirements for reports on satellite construction, on any tests that show substandard performance, and on satellite launches and performance in orbit. Further, there are requirements for reports that indicate how much of the satellite is utilized and for what purpose, and even semiannual reports on specific test measurements indicating the technical and operational aspects of each satellite network, as well as detailed frequency and carrier plans.[34]

In a competitive national satellite market all of these requirements can be difficult, not only in terms of providing the right response or information in a timely manner, but also in not disclosing data or intelligence that could be used in a damaging way by existing or would be competitors. The "open skies" policy of the United States has stimulated new services, multiple satellite systems that provide very cost-effective offerings to consumers, but it has made the regulatory process on many occasions rather cumbersome for government officials and the commercial operators of satellite services. In the context of international satellite services it has made the United States a rather difficult international partner to deal with within INTELSAT and INMARSAT, but on balance the advantages have seemingly outweighed the disadvantages.

5.3.6 Other Regulatory Satellite Issues

There are many other national regulatory issues that apply to seeking and maintaining a license for a satellite system, although the more important aspects have been summarized above. Further, the national procedures are only the first step. The international procedures for frequency registration and of intersystem coordination as discussed in the previous section (5.3.3) can be a long and painstaking prelude to the ultimate registration of a satellite system in the Master Frequency Registry of the ITU.

There remains yet another consideration in this process, this being the concerns of other countries with regard to the operation of a satellite system within their own territories if the system is global, regional, or even involves transborder services on a largely domestic system. In short, simply because a system has obtained a national license for frequencies and orbital locations for new satellite networks and then fulfilled the ITU procedures for intersystem coordination and registration of networks on the Master Frequency Registry does not mean all requirements are satisfied. One next must obtain national operating licenses and landing rights to operate international satellite services in a particular country. Each country that is involved in the process still has their own separate regulatory controls. As noted earlier, there is the matter of the earth station approvals in the affected countries and then the even more difficult landing rights and operating license.

Most countries regulate the number, qualifications, and characteristics of international communications carriers who are authorized to operate within their territories. In some countries the operator must have full or partial local ownership. In the United States there are rather streamlined procedures for registering for such status. Formal action by the FCC itself is not even required since these procedures are implemented by the International Bureau. In most other countries, however, new international communications are approved rather reluctantly and sometimes only after long delays, if they are approved at all. In Europe, there has been positive pressure from the European Commission, which favors competitive access in the satellite and the telecommunications arena, but at the national level new "landing rights" agreements have been difficult to obtain in a timely manner.

This is not to say that legitimate national regulatory concerns are not involved. Key questions involve such issues as unauthorized transborder data flow of information of a strategic or sensitive nature and the offering of private versus public services. Private carriage of telecommunications can, over time, undercut the cost efficiency of the public service. Local authorities have questions as to the appropriate technical, operational, and tariffing policies concerning the interconnection to the national public switched telecommunications network. This usually leads to the very sensitive issue of the bypassing of national telecommunications services and to what extent this may be considered inappropriate "cream skimming" under local regulatory controls and laws. In these and other matters such a "footprint spill-over" into other countries, the most common practice has simply been one of direct negotiations with the national regulatory officials involved.

In cases where negotiations are slow or unproductive, U.S. governmental trade officials and officials of the European Commission, especially Directorates Number IV, XII, and XIII (i.e., those Directorates that deal with competitive practices, telecommunications, and space) have sometimes intervened to encourage the discussions forward. The regulatory landscape for satellite communications is both diverse and complex. The control points are hard to identify in a complete and comprehensive manner. Just when the inventory of regulatory procedures seems complete, another regulation or another player always seems to be added.[36]

5.4 COMPARATIVE ASSESSMENT OF DIFFERENT TYPES OF SATELLITE SYSTEMS

The advent of numerous types of satellite configurations has added great complexity to understanding what the proper technical, operational, and financial comparisons are that can be made about these various systems. Table 5.2 seeks to make some comparison among the space segments. It should be noted that this does not seek to address ground segment costs where the GEO systems would tend to require a more expensive investment in earth station facilities.[37]

In this comparison, two key aspects emerge. In terms of cost-effective coverage of large areas the geosynchronous satellite is still likely to be the winner among satellite options. This conclusion must be offset against other factors such as the following: (a) the need for improved look angles and much smaller terminal equipment for mobile satellite services; (b) the improved opportunities for spectrum efficiency through higher levels of frequency reuse achievable by lower orbiting satellites; and (c) the reduced latency or delay in the transmission path. The recent Hughes Spaceway filing perhaps best typifies the most efficient geosynchronous concept, while Teledesic represents the most advanced concept for LEO systems.

The second observation is that there is a very new type of telecommunications platform that demonstrates a remarkable capability in terms of broad coverage, cost-effective service, and the ability to work with mobile and low-cost fixed terminals. One will note that the concept known as the High Altitude Long Endurance (HALE) platform seems to outperform the various satellite configurations. This HALE platform can be a free-flying or virtually geostationary unit equipped with telecommunications, remote sensing or other payloads typically operating at 18.5 kilometers or above. These HALE platforms operate in a new domain that has been called "proto-space." Proto-space has been suggested as that region between 18.5 and 110 kilometers in altitudes that are above normal commercial aviation space and below the known practical applications or research purposes of outer space systems. Today outer space is defined as beginning at 110 kilometers and the lowest known satellite ever practically orbited was the Skynet satellite, which was launched in a nearly 100 kilometer orbit. In the future there appear to be a number of practical applications for proto-space that can now be anticipated and therefore some appropriate means of addressing this sector would seem highly useful. This would include identifying the applications, allocating frequencies, and other key policy or regulatory issues. The two known applications of proto-space at this time include operation of various types of HALE platforms and

System Type	System Cost ($US)	Coverage (km²)	Beam Throughput	Beam Performance Index
GEO (3 satellites) (8 beams)(7-yr life)	$1.2 B	12.6×10^6	200 Mb	$2834/mb/km²/yr
MEO (12 satellites) (20 beams)(5-yr life)	$2 B	3.1×10^6	250 Mb	$6451/mb/km²/yr
LEO (50 satellites) (40 beams)(5-yr life)	$4 B	1.5×10^6	50 Mb	$8680/mb/km²/yr
MEGA-LEO (800 satellites) (50 beams)(5-yr life)	$12 B	1.5×10^6	100 Mb	$6000/mb/km²/yr
HALO (12 cells) (10-yr life)	$10 M	0.18×10^6	200 Mb	$1852/mb/km²/yr
TERRESTRIAL (30 cells)(20-yr life)	$50 M	0.008×10^6	90 Mb	$82,000/mb/km²/yr
Sample Calculation—GEO System:				

Costs ($1.2 Billion)

(200 Mb) × Beam size (12.6 m.km) × satellites (3) × beams (8) × life (7-yr) × (eff) 100 percent

Resulting Beam Performance Index: $2834/Mb/km²/yr.

Note: Efficiency index for LEOs and MEOs is 30 percent because of ocean and arctic coverage 70 percent of the time.

TABLE 5.2   COMPARING DIFFERENT SATELLITE SYSTEM CAPABILITIES

the operation of so-called space planes for research, and eventually, for space tourism. HALE platforms would tend to operate in the 18–21 kilometer altitude region and space planes would typically operate in altitudes of up to 60 kilometers or more.

The HALE platform as now conceived would involve one or more types of propulsion systems that include high efficiency turbines with 10-day duration missions, fuel-cell and solar-cell-powered systems (sometimes called the eternal airplane but, in fact, capable of about 120-day sustained missions), and microwave beamed energy from the ground to sustain very long duration missions indeed. Typically, a HALE platform would involve two unmanned, robotically controlled units that would alternate with one another for the purposes of maintenance, repair, retrofit, or payload upgrade. The initial price estimates for a HALE platform would be on the order of $10 million for a two-unit system or $5 million per typical proto-space plane. These platforms cannot only provide a wide range of telecommunications services but also navigational services, remote sensing, news gathering, or even environmental data collection. The advent of HALE platforms in so-called proto-space will likely redefine the relationship between space communications and terrestrial services with the creation of a new transitional service between these two sectors.[38]

5.5 SATELLITE VERSUS TERRESTRIAL TELECOMMUNICATIONS STANDARDS

The process of filings, construction permits, licensing, comparative hearings, and registration of satellites and antennas as discussed in the previous sections are all regulatory controls by governments to supervise telecommunications satellites in order to meet what might be called social objectives. These objectives include prevention of overlap, undue interference to space and terrestrial communications systems, "cream-skimming," and fraudulent behavior, saturation of the radio frequencies, discriminatory pricing, and unfair market advantages. These objectives can also have a positive dimension such as promoting the lowest-cost pricing, educational and health related development, national security, and disaster warning and management services. These regulatory procedures are to be clearly distinguished from the much different role of technical standards. Standards, which are created by national and international technical bodies, have as their objective, the following: (a) interoperability; (b) uniformity and compatibility; (c) economy of scale; (d) certification of minimum performance and technical quality; (e) health standards; and (f) elimination of higher-cost and noncompatible proprietary systems.

The entities that make the key international standards for satellite communications are the International Telecommunication Union, the International Electro-Technical Commission, the International Standards Organization, and the IEEE. To a certain extent the major international satellite organizations such as INTELSAT, INMARSAT, and EUTELSAT also contribute to this process. Finally the regional groups such as the European Telecommunications Standards Institute (ETSI), the Council on European Telecommunications (CEPT), the Organization of American States and its CITEL and COMCITEL Committees, the Asia-Pacific Telecommunity, and others also contribute to the satellite standards-making process.[38]

The field of satellites, when it comes to standards-making, suffers from certain disadvantages. This is a result of the historical fact that the international standards for terrestrial telecommunications have for many years been developed within the ITU Consultative Committee on International Telephone and Telegraph (CCITT), while the standards for satellites have been essentially developed within the Consultative Committee on International Radio. Since many of the key new digital standards such as those for the Integrated Services Digital Network (ISDN) have been developed within the CCITT (e.g., CCITT Study Group 18) the satellite standards often tend to be added-on afterthoughts or derivative from the terrestrial standards. In some cases this works quite well and in other cases, satellites can be placed at some disadvantage. Fortunately, the ITU is being restructured to combine terrestrial wire and wireless technologies together in terms of standards-making. This makes a great deal of sense because hybrid wire and wireless systems that are linked together to provide integrated services will be increasingly common in the twenty-first century. This does not necessarily solve current "inconsistent standards" involving wire and satellite services.

Although there are many examples of this type of "inconsistency" among and between different transmission media, a few examples will be provided to illustrate the different types of issues that are involved. The traditional wire-based terrestrial telecommunications systems is premised upon the concept of traffic concentration. Traffic is routed through local wires into a switch that, in turn, directs traffic to large switches or transit centers and perhaps ultimately to a national gateway where international switches are located. After traveling by submarine cable to another country, the traffic is then deconcentrated through multiple switches back to the end-user. This basic telecommunications architecture is sometimes known as an hierarchical or vertical communications system. This type of network architecture was illustrated in Figure 1.9 in Chapter 1.

By contrast, satellites that have on-board switching capability over a large service area can "bypass" the concentrators and instead link end-users to end-users directly without transiting four to ten different switches. The type of model that is used for telecommunications transmission standards should in theory be different from wire- and satellite-based systems. In the first or traditional model , namely the hierarchical approach, the switches are predominant in terms of noise and interference allocations. In the horizontal or distributed end-user-oriented model, which applies to the most modern satellites (as presented in Figure 1.11 in Chapter 1) the transmission medium is the key. The ITU approved "universal model" is called a hypothetical reference connection (HRX). This model features the hierarchically switched wired-based concept and satellites are simply required to conform. Time and time again in areas such as transmission delay, quality standards for digital bit-error rates, hypothetical reference connections (HRXs), and many other areas, the ITU standards tend to position satellites as the exception rather than as another important rule. Fortunately with advanced error-correction techniques, increased on-board intelligence, and improved technologies, satellites have been largely able to adapt. The single biggest problem for the future is the increasing pattern of the integration of wire and wireless systems and the lack of clear standards to address the special problem of hybrid wire and satellite systems, which in turn may be linked to digital networks with their own processing requirements. In general the networks of the future will likely evolve to a dynamic balance between horizontal and vertical that tend to be like a modified "mesh" network, as earlier shown in Figure 1.12. Clearly, there are basic differences between wire and wireless networks that will impact heavily on future standards-making activities. These distinctions are reflected in Figures 5.9 and 5.10, which depict the difference between the "spider webs" of terrestrial nets and the "sunshine" architecture of wireless networks, and especially, of satellites.[39]

The unplanned, unstructured, and nonstandardized linkage of wire-, wireless-, satellite-, and computer-based LAN networks may well produce cumulative effects that are harmful to high-quality reliable telecommunications. Clearly, recent trends to integrate wire and wireless planning and standards-setting are a step in the right direction. This could ultimately produce new and better standards covering all forms of digital transmission systems for the twenty-first century.

5.6 CONCLUSIONS

The potential for satellites to provide a wide range of digital telecommunications services for many decades to come should not be underestimated. Today, there is much interest and enthusiasm in fiber optic transmission systems, optical switching, and photonic-based information services. This is natural and appropriate particularly given the impressive gains

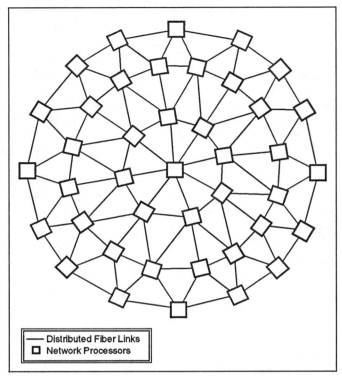

FIGURE 5.9   The "Spider Web" Network of Terrestrial Wireline Systems

being achieved in advanced areas like soliton pulse fiber systems, plans for multigigabit optical processors and switches, color modulation (i.e., dense Wave Division Multiplexing) systems, and lower-cost optical systems.

This does not in any way suggest that satellites for many types of unique services are somehow obsolete. It is a question of "and" rather than "either/or" technologies. In the areas of broadcasting, mobile services, radio navigation, and determination, satellites will continue to play important roles. Furthermore, new technologies such as on-board signal regeneration, on-board signaling, Josephson Junctions, and High Altitude Long Endurance platforms could help major satellite and proto-space advances in coming decades.[40]

The key elements to emphasize with regard to satellite communications are that a complex set of laws, regulations, international recommendations, approvals, and standards apply and that local, regional, national, and international constraints are all involved. Important considerations include the following:

(a) Both earth station antennas and satellite facilities are subject to controls.

(b) In the case of earth station antennas, two-way facilities are usually subject to more stringent controls.

(c) In many countries the construction of either an antenna or a satellite system requires a construction permit and an operating license. (In a number of countries where lib-

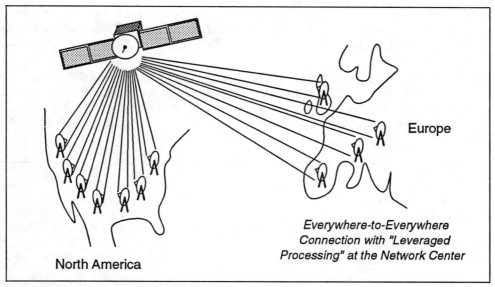

FIGURE 5.10 The "Sunshine" Network of Wireless and Satellite Systems

eralization has not occurred only the military or the government telecommunications agency can own antennas or satellites, but increasingly these restrictions are being relaxed around the world.)

(d) Since there are a limited number of frequency bands authorized for satellites, the licensing of a new satellite system through national and then ITU procedures is long and rather complex.

(e) Most of the frequencies available for satellite expansion are in the much higher frequency bands such as 30 GHz/20 GHz, which are only now beginning to be exploited or in the bands newly allocated by the ITU to low and medium earth orbit satellites for mobile communications.

(f) There are many legal, quasi-legal, and international policy issues related to satellite communications. These include transnational data flow, spill over of beam signals across national boundaries, national sovereignty claims to outer space and orbital arcs, "country of convenience" satellite filings, etc.

(g) Local jurisdictions also can regulate earth station siting, pointing characteristics, power levels, and operation for reasons of health and zoning.

(h) The many different types of satellite orbits, satellite services, frequency bands, geographic coverages, and technical and operational characteristics make it difficult to regulate and control satellite systems on an equitable and consistent basis.

(i) The standards under which satellite services are provided, especially PSTN services, are sometimes a major challenge for satellites, often because the basic standard was first developed for terrestrial wire technology. (The even greater challenge for the future may be for integrated wire and wireless systems, which must operate as hybrid networks.)

(j) Apart from international regulatory controls and standards for satellite systems and earth station antennas, there are other issues that operators of international satellites systems must address in the form of "landing rights" and interconnection to local PSTN services.

(k) Despite all of the barriers and regulatory hurdles needed to establish and operate a satellite system, it is still an attractive and potentially profitable business, which was equivalent to $11.4 billion in global revenues in 1992, was about $14 billion in 1994, and will be close to $40 billion worldwide in the early twenty-first century. See Table 5.3.[41]

Satellite Service	1992	2002
Fixed Satellite Services • INTELSAT • Regional and Other International Satellite Systems • U.S./Canada National Systems • Other National Systems	$4.5 Billion $1.8 Billion $2.3 Billion $1.4 Billion	$8.5 Billion $3.6 Billion $4.5 Billion $3.4 Billion
Fixed Satellite Service (Total)	$10.0 Billion	$20.0 Billion
Mobile/Low Orbit Services	$0.8 Billion	$10.0 Billion
Broadcast Satellite Services	$0.5 Billion	$8.0 Billion
Military Satellite Services	*	*
Other (e.g., Data Relay, etc.)	$0.1 Billion	$0.3 Billion
Total Services	$11.4 Billion	$38.3 Billion
*No accurate or meaningful figures for military services are readily available.		

TABLE 5.3   FUTURE PROJECTION OF GLOBAL SATELLITE MARKETS

Despite all of the technical challenges posed by fiber optic cables, HALE platforms, and other systems, the future of satellite communications is still likely to be a bright one. New technical developments, new system architectures, and well-focused research and development programs (especially in Europe and Japan) are likely to pay good dividends in the next decade. Despite the importance and dynamism of satellite technology, it still seems probable that liberalization and regulatory reform will have the greatest impact on the growth of satellite systems, and particularly on totally new ECO, LEO, and MEO satellite systems.[42]

ENDNOTES

(1) Howkins, J., and J.N. Pelton, *Satellites International* (London: Macmillan Press, Ltd., 1989).

(2) "ITU International Table of Frequency Allocations," as contained in Part 2 of the FCC Rules and Regulations (Washington, DC: U.S. Government Printing Office, 1994).

(3) Meagher, C., *Satellite Regulatory Compendium* (Potomac, MD: Phillips Publishing, 1993), pp. 2.1-2.30.

(4) "The Green Paper on Satellite Communications" (Brussels, Belgium: European Commission, 1992).

(5) "The INTELSAT Agreement and Operating Agreement," TIAS Series (Washington, DC: U.S. State Department, 1992); also see "Comsat Appeals for Privatization," *Satellite Communications*, April 1994, pp. 10-11; also see "How INTELSAT Was Privatized While No One Was Looking," *Via Satellite*, February 1989.

(6) Payne, S., ed., *International Satellite Directory* (Sonoma, CA: Design Publishers, 1994).

(7) Pelton, J.N., *The How To Book of Satellite Communications* (Sonoma, CA: Design Publishers, 1992), pp. 18-65.

(8) Meagher, C., *Satellite Regulatory Compendium* (Potomac, MD: Phillips Publishing, 1993), pp. 2.31-2.160.

(9) ——, *Satellite Regulatory Compendium* (Potomac, MD: Phillips Publishing, 1993), pp. 2.36-2.90.

(10) "The PACSTAR Satellite System," FCC Filing, Common Carrier Bureau, 1985.

(11) "Tongasat," *Via Satellite*, November 1991, p. 14.

(12) "Reformed ITU Filing Procedures—Brokers for Orbital Space-Boon or Bane?" *Satellite Communications*, May 1994.

(13) "Ku-Band Sharing," *Satellite Communications*, April 1994, p. 12.

(14) "Band Sharing, 6; Band Splitting 1: Big LEO Report Sent to FCC," *Signals*, Spring/Summer 1993, No. 6, pp. 1-4.

(15) Motorola's Pioneer Preference Application for Iridium Loses," *FCC Week*, April 1992, p. 1.

(16) Motorola's Pioneer Preference Application for Iridium Loses," *FCC Week*, April 1992, pp. 2-3.

(17) "Crowding of the Communications Satellite Bands," *Via Satellite*, May 1993, p. 18.

(18) Pelton, J.N., *The How To Book of Satellite Communications* (Sonoma, CA: Design Publishers, 1992), pp. 87-110.

(19) "The INTELSAT Agreement and Operating Agreement," TIAS Series (Washington, DC: U.S. State Department, 1992), Article XIV, p. 39.

(20) Lewis, G., *Communications Services Via Satellite*, 2nd ed. (Oxford, U.K.: Butterworth-Heinnemann Press, 1988).

(21) "The INTELSAT Agreement and Operating Agreement," TIAS Series (Washington, DC: U.S. State Department, 1992), Article XIV, pp. 39-40.

(22) "Reformed ITU Filing Procedures—Brokers for Orbital Space-Boon or Bane?" *Satellite Communications*, May 1994, p. 12.

(23) Lewis, G., *Communications Services Via Satellite*, 2nd ed. (Oxford, U.K.: Butterworth-Heinnemann Press, 1988), pp. 69-70.

(24) Pelton, J.N., *Global Communications Satellite Policy: INTELSAT, Politics and Functionalism* (Mt. Airy, MD: Lomond Systems, 1974).

(25) "The VSAT Market," *Satellite Communications*, April 1994, p. 16; also see Everett, J., *VSATs* (London, U.K.: IEEE, 1992).

(26) Meagher, C., *Satellite Regulatory Compendium* (Potomac, MD: Phillips Publishing, 1993), pp. 2.93-2.104; also see 74 FCC 2D 205 (1979), "Regulation of Domestic Receive-Only Satellite Earth Stations."

(27) ——, *Satellite Regulatory Compendium* (Potomac, MD: Phillips Publishing, 1993), pp. 2.100-2.119.

(28) FCC Document, "Processing Procedures and Attachments" (75-932), pp. 12-31.

(29) FCC Document, "Processing Procedures and Attachments" (75-932), pp. 16-20.

(30) FCC Document, 35 FCC 2D "DOMSAT II."

(31) FCC Document, 70 FCC 2D 1853, 1979; also see FCC Document, 47 USC 214, 1983.

(32) "The Green Paper on Satellite Communications" (Brussels, Belgium: European Commission, 1992).

(33) "FCC Rules of Procedure and Processing Orders," Rule 25 (1990).

(34) Meagher, C., *Satellite Regulatory Compendium* (Potomac, MD: Phillips Publishing, 1993), p. 2.100-2.104.

(35) Verhoff, P., and D. Wright, *Journal of Space Communications*, Special Edition on the Regulation of Satellite Communications (Amsterdam, Netherlands: IOS Press, September 1992), pp. 2-13.

(36) Bloomfield, R., "International Telecommunications Standards-Making,"Lecture, April 22, 1994, University of Colorado.

(37) Pelton, J.N., "Overview of World Satellite Systems: Key Trends and Patterns," 45th Congress International Astronautical Federation, Jerusalem, Israel, October 1994.

(38) Maul, J., "High Altitude Long-Endurance Aircraft," Workshop on Wireless Power Transmission, Department of Commerce, January 31, 1994, pp. 33-73; also see Brown, S., "The Eternal Airplane," *Popular Science*, April 1994, pp. 1ff; also see Glaser, P.E., "The Power Relay Satellite," Plenary Lecture, 44th Congress International Astronautical Federation, Graz, Austria, October 16-22, 1993.

(39) Pelton, J.N., "Five Reasons Why Nicolas Negroponte Is Wrong About the Future of Telecommunications," *Telecommunications*, April 1993.

(40) *Communications Satellite Technology and Systems*, NASA and NSF Report (Baltimore, MD: International Technology Review Incorporated, 1993).

(41) Pelton, J.N., "Future Trends in Satellite Communications," NEC ComForum, Orlando, FL, November 1992.

(42) Pelton, J.N., "Overview of World Satellite Systems: Key Trends and Patterns," 45th Congress International Astronautical Federation, Jerusalem, Israel, October 1994, pp. 2-8.

PART III

SPECTRA, STANDARDS, POLICY AND REGULATION

CHAPTER SIX

THE ELECTROMAGNETIC SPECTRUM AND FREQUENCY ALLOCATIONS FOR WIRELESS AND SATELLITE SERVICE

6.1 INTRODUCTION

The issue of how energy could be transmitted over a long distance has been a subject of research for many centuries. Ever since the time of James Clerk Maxwell in the nineteenth century it has been understood that radiated energy of all types follow the "rules" of electromagnetism. In short, whether irradiated energy is electric current, heat, various forms of radio waves, infrared, visible light, ultraviolet, X-rays, gamma rays, or cosmic rays, it is essentially a similar phenomena albeit at different wavelengths, frequencies, and energy levels. In general the higher the frequency and the more focused the beam, the higher the energy level.[1]

Leaving aside the so-called "particle-like" characteristics of electromagnetic radiation, the most straightforward way to view telecommunications applications and their use of the electromagnetic spectrum is in terms of wave forms. Certain basic mathematical rules apply. Most important of these rules is that the frequency of the radiated signal is always traveling at the speed of light or the wave-form velocity divided by its wavelength. Thus frequency increases as the wavelength becomes smaller and smaller. The complete electromagnetic spectrum is provided in Table 6.1.[2]

Electromagnetic waves can be modulated for the purposes of sending telecommunications signals. First, this was done in analog form using techniques such as amplitude modulation (AM), frequency modulation (FM), or single side band (SSB) modulation. Subsequently, digitally based multiplexing techniques were used, such as Timed Division Multiple Access (TDMA) and most recently, Code Division Multiple Access (CDMA), which allow greatly enhanced performance in terms of derived channels per hertz of bandwidth.[3]

Name	Frequency Range	Wavelength
Low Frequency (LF)	10–300 kHz	30–1 km
Medium Frequency (MF)	300–3000 kHz	1000–100 m
High Frequency (HF)	3–30 mHz	100–10 m
Very High Frequency (VHF)	30–300 mHz	10–1 m
Ultra High Frequency (UHF)	300–3000 mHz	100–10 cm
Super High Frequency (SHF) (Microwave Band)	3–30 gHz	10–1 cm
Extremely High Frequency (EHF) (Millimeter Wave Band)	30–300 gHz	10–1 mm
(*Note*: Above these bands one finds infra-red, visible light, ultraviolet, X-rays, gamma rays, and cosmic rays.)		

TABLE 6.1   THE ELECTROMAGNETIC SPECTRUM

Several general characteristics of the electromagnetic spectrum are worthy of special note. First, as one moves up the electromagnetic spectrum certain problems occur at higher and higher radio frequencies. One problem is that the smaller waves do not "bend" around obstacles as do the longer wavelengths. This means that the signal needs to either have a direct line-of-sight between sender and receiver or one must rely on reflections of the signals called "multipath" to reach the wanted receiver location. Second, the higher frequencies are increasingly sensitive to atmospheric distortions or absorption. These reductions in signal can come from rain, or what is more precisely called "precipitation attenuation." Distortions or even loss of signals can also come from heat scintillations, or atmospheric absorption that occurs particularly in the higher frequency bands such as the microwave band or even more so in the millimeter wave frequencies. This means that high-power and more directional (or higher-gain) antennas must be used to overcome these effects.[4]

A third issue is that more precise oscillators and more precisely shaped antennas must be developed to operate at these higher frequencies. At higher frequencies the antenna reflector can be smaller but the precise shaping and forming of the beam must also increase. Higher frequency means smaller waves. A fixed antenna aperture size allows more "wavelengths" in the transmitted or received signal to be generated or captured as the frequency goes up or the wavelength goes down. In this respect the gain of an antenna is inversely proportional to the square of the wavelength. As the need for frequency reuse increases there will also be a need for more and more carefully shaped beams to avoid the problem of so-called "hot spots," "dead spots," and cell boundary overlap. Given these three difficulties, one might question the wisdom of using the higher frequencies at all.[5]

There is, however, one overriding advantage associated with the higher frequencies that makes these difficulties seem worthwhile. This clear-cut plus is quite simply the increased spectra or band-width that is available. Every time frequencies increase ten times, the available frequency band-width correspondingly increases ten times as well. Take the example of a television channel that now conventionally requires some 6 MHz of spectra to carry all the information needed to create a clear color image. If one were to attempt to use, say, the MF or HF bands for television service, this would make very little sense. This is because the television channels would not "fit" or they would quickly consume the entire band to the exclusion of all other uses. In short, broad band services imply the use of the upper frequencies and the need to access very broad segments of the spectrum.

To continue the illustration, television was first allocated frequencies in the VHF band. When those were exhausted, there were additional allocations made in the UHF bands as well. Today those allocations are, in effect, "grandfathered" into practice by decades of use, but they are considered today to be a wasteful use of this valuable spectra. In future years one can anticipate efforts to shift television services, at least in part, to the higher millimeter wave bands.

This impetus to shift to higher frequencies is based on at least two reasons. First, cable television systems can typically provide television services to the home by coaxial cable or fiber without any "free space" or environmental use of spectra. Second, the much higher frequencies in the millimeter wave band could be used for this purpose with room to spare for lots of new TV channels and HDTV channels as well. New systems operating in perhaps the 28–30 GHz band or even higher (i.e., 42–44 GHz in the United Kingdom is being allocated to this purpose) would allow 50–100 TV channels of capacity to be achieved. This approach could also use cellular techniques to limit power plus digital compression techniques to increase spectrum efficiency. The special appeal and cost efficiency of digital efficiency is almost a third reason for contemplating the migration to higher frequencies.

Some believe that these reasons are strong enough, at least in the United States, for the FCC to opt to migrate over-the-air television broadcasting over a period of years from VHF and UHF up to SHF and EHF frequencies. This action would free spectra for the new high-demand mobile communications services such as PCS. Such matters are never simple, however, since both terrestrial and space communications users are contesting for access to the higher frequencies as well.[6]

Although this is only one example, it is nevertheless indicative of the trends currently at work in U.S. public policy with regard to frequency allocations and spectra management. In general, these trends will likely follow these policy and regulatory guidelines.

Further Reforms in the Use of the Frequency Band.   There will be reforms and procedural revisions to the process of frequency spectrum allocations. This is because spectra is today not necessarily well allocated nor is it being efficiently used.

This is a pervasive problem that extends to private telecommunications users, but especially to governmental and military uses. The causes of today's problems include exploding consumer demand, new technological developments and innovations that require expanded frequency spectra, outmoded and inefficient patterns of historical planning and use, major inefficiencies in governmental and military usage, as well as the obsolescence of some telecommunications technologies and services.

Economic Optimization.   Spectra use will be increasingly optimized using economic value and economic return measures of efficiency. Radio band-width is thus of ever-increasing value as a scarce resource. The increasingly competitive demands for the use of the spectrum for a variety of telecommunications purposes will likely continue to increase. Lower-value users will thus be gradually replaced by higher-value users and operators.

Governmental Policy Oversight.   This will most likely tend over the next decade toward the auctioning off of spectra to competitive bidders as it is reclaimed from less efficient uses of band-width. In fact, after years of debate and political partisanship on this point, the 1992 Omnibus Budget Bill mandated that the FCC begin the auctioning of frequencies as soon as possible, starting with the new allocations for Personal Communications Services. The discrepancy between the mobile, cellular, ESMR, and PCS industries paying billions of dollars for their spectra and radio and television operators obtaining their allocations for free cannot continue. Furthermore, the U.S. approach of auctioning off frequencies in their internal or national "market" while still seeking "free allocations" for, say, international satellite communications is another potential problem. This inconsistency or at least the perceived inconsistency will likely become a practical difficulty in the not-too-distant future, as noted below.[7]

Phase Out of Underutilized Spectrum Allocations.   The clearly underutilized allocations will be phased out of service and then reallocated in a competitive bidding process to much higher-value users. This means reallocating use among private and commercial users but also implies the shift of frequencies used by the federal government and the military. The clear trend is shown by the recent experience with the auction of narrow band PCS frequencies in mid-1994 and then the broad band PCS frequency auction in December 1994. These two auctions netted the U.S. government billions of dollars. This first auction, of a total of some 123 MHz of capacity, clearly set the stage for much more activity of this sort within the United States, but probably internationally as well. In fact, the first country to auction frequencies was actually New Zealand in the early 1990s. Their approach of awarding frequencies to the "second-highest" bidder, however, involved a rather odd twist to this auctioning process.[8]

The Changing International Frequency Allocation Process.   Finally there is the issue of international frequency allocations and new patterns of use. The U.S. shift to the auctioning of licenses for frequency use will undoubtedly lead to formal changes at the World Administrative Radio Conference, which will be held in 1995. Here the fundamental issue may well be the move by a number of countries, especially from developing economies to back the idea of receiving compensation for the use of frequencies for such applications as international satellite communications. In short, if the process of national frequency auctions works and raises large sums of money why not use a similar system in the world community? It is certainly difficult to declare that this process increases economic efficiency and technological progress at the domestic level and then argue the reverse at the global level. Nevertheless, there are myriad issues that will not be easily resolved. The value of frequencies greatly changes from country to country. There is a great variation in market demand from country to country. There are serious questions as to the value and "ownership" of frequencies over the high seas and polar caps. There is also the question of what the boundary lines for frequencies are in adjacent countries and there are even questions of enforcement procedures in case of disputes.[9]

Coping with Globalism.   The first uses of telecommunications beginning with the telegraph and the telephone were essentially local. Over time, however, service was extended overseas and telecommunications became international. The technology and the demand was, however, of a nature that it could be controlled through regulated entry and exit points at the nation's borders. Today as telecommunications services become increasingly digital, broad band, and in many cases wireless, the element of control has broken down in terms of DBS spill-over, wireless radio telephone roaming, and VSAT customer-premise terminal networks. The trends for the future suggest that global wireless and international roaming will become even more widespread. This creates issues of governmental control on one hand and the increasing need for globally accepted rather than regionally or nationally decided allocations of frequencies. If the future projected in Figure 6.1 indeed becomes reality, then the feasibility of having national or regional frequency allocations will become increasingly untenable.

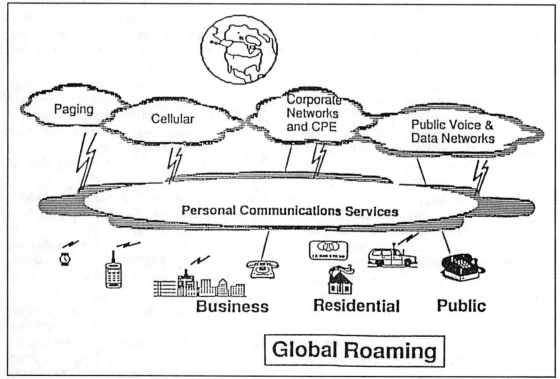

FIGURE 6.1   Integrated Wireless Services via a PCS Network

6.2 KEY FREQUENCY ALLOCATIONS

Before one can begin to enhance the frequency allocation process to make it more equitable and efficient, it is important to understand where the current baseline is. This is complicated by the fact that there is both an international and national process at work. The international frequency allocations are those that are agreed upon by all the nations who participate with full rights of a treaty-making body in the ITU World Administrative Radio Conference Final Acts. Then after the international process is completed there are the actual allocations, which have been made at the national level unless they involve a global or regional service.

The patterns of use around the world are, in fact, greatly different. Many developing countries are today, for instance, still using high frequency or short-wave bands for telephone services, while in other countries of the OECD, for instance, these bands are used instead for television or mobile communications. The chart in Table 6.2 provides a condensed summary of the Region 2 or "Americas" frequency band allocations for just the HF frequencies as provided in the chart and just for the major service categories. As noted earlier, this regional approach may very well be increasingly incompatible with both emerging wireless telecommunications technology and the various applications that look to international roaming and global connectivity.[10]

It should be noted that while Table 6.2 accurately portrays the complexity and intricacy of the frequency allocations in the high frequency band, this is in some sense just the tip of the iceberg. The shared bands have primary and secondary users. There are footnotes whereby some countries within the region have taken exception to the allocations. There are in some cases even up to four shared allocations in a particular band.

Furthermore, the allocations shown in Table 6.2 are far from completely comprehensive. There are at least some allocations in the HF band for such additional services as amateur satellite, radio astronomy, space research, standard frequency and time signals, broadcast satellite, radio location, land mobile satellite, earth exploration satellite and mobile satellite services. The unique capabilities of different frequency bands make them desirable for a very wide range of applications and the ITU allocation procedures have tried very hard to accommodate everyone by giving everyone a sliver of the pie.[11]

This is often more accommodating than efficient. Every different allocation tends to require guard bands between the services which in effect ends up as "wasted" spectra. Furthermore with primary, secondary, and even tertiary and quadranary users, the steps required to prevent interference and crosstalk can lead to extraordinary measures that are expensive and technically demanding. This can range from special filters and polarization techniques, to special modulation and encoding approaches, or enhanced antenna sidelobe designs.

More recently the strategy has shifted to one of trying to move appropriate applications and services to higher bands with larger spectra and less segmentation in the allocations. This process has, at least in the United States, also moved in the direction of shifting users to different bands where possible and letting the economic gains generated by these shifts actually pay for the transition process. In the longer term this should result in overall "win–win" results for all of the concerned users. Some of the current strategies designed to help

Aeronautical/Aeronautical Mobile Bands		Fixed Bands	
3.000– 3.155	MHz	3.155– 3.400 (Shared)	MHz
3.400– 3.500	MHz	4.000– 4.063 (Shared)	MHz
4.650– 4.750	MHz	4.438– 4.650 (Shared)	MHz
5.450– 5.730	MHz	4.750– 4.995 (Shared)	MHz
6.525– 6.765	MHz	5.005– 5.450 (Shared)	MHz
8.815– 9.040	MHz	5.730– 5.950 (Shared)	MHz
10.100–10.150	MHz	6.765– 7.000 (Shared)	MHz
11.175–11.400	MHz	7.300– 8.195 (Shared)	MHz
13.200–13.360	MHz	9.040– 9.500 (Shared)	MHz
15.010–15.100	MHz	10.150–11.175 (Shared)	MHz
17.900–18.030	MHz	12.050–12.230	MHz
21.924–22.000	MHz	13.410–13.600 (Shared)	MHz
23.350–23.400	MHz	13.800–14.000 (Shared)	MHz
		14.350–14.990 (Shared)	MHz
		15.600–16.360	MHz
Amateur Radio Bands		17.410–17.550	MHz
		18.068–19.680 (Shared)	MHz
3.500– 4.000	MHz	19.800–19.990	MHz
7.000– 7.300	MHz	20.010–21.000 (Shared)	MHz
10.100–10.015	MHz	21.850–21.924 (Shared)	MHz
14.000–14.350	MHz	22.855–23.350 (Shared)	MHz
18.068–18.168	MHz	23.750–24.990 (Shared)	MHz
21.000–21.450	MHz	26.480–28.000 (Shared)	MHz
24.890–24.990	MHz	29.800–30.000 (Shared)	MHz
28.000–29.700	MHz		

Broadcasting Bands		Maritime Mobile Bands	
5.950– 6.200	MHz	4.000– 4.063 (Shared)	MHz
9.500– 9.900	MHz	4.063– 4.438	MHz
11.650–12.050	MHz	6.200– 6.525	MHz
15.100–15.600	MHz	8.195– 8.815	MHz
17.550–17.900	MHz	12.230–13.200	MHz
21.450–21.850	MHz	16.360–17.410	MHz
25.670–26.100	MHz	18.780–18.900	MHz
		19.680–19.900	MHz

Mobile Band		Land Mobile Bands	
3.155– 3.400 (Shared)	MHz		
4.438– 4.650 (Shared)	MHz	25.010–25.070	MHz
4.750– 4.950 (Shared)	MHz	25.210–25.330	MHz
5.060– 5.450 (Shared)	MHz	26.175–26.480	MHz
5.730– 5.950 (Shared)	MHz	27.410–27.540	MHz
6.765– 7.000 (Shared)	MHz	29.700–29.800	MHz
7.300– 8.100 (Shared)	MHz		
10.150–11.175 (Shared)	MHz		
13.410–13.600 (Shared)	MHz		
13.800–14.000 (Shared)	MHz		
14.350–14.990 (Shared)	MHz		
20.010–21.000 (Shared)	MHz		
23.000–24.890 (Shared)	MHz		
26.480–28.000 (Shared)	MHz		
29.890–29.910 (Shared)	MHz		

TABLE 6.2 Key Frequency Allocations—3 to 30 GHz

address this problem of historical patterns of frequency allocations and the need for greater spectra efficiency are discussed in more detail in the next section.[12]

6.3 STRATEGIES FOR COPING WITH FREQUENCY SHORTAGES IN WIRELESS SERVICES

In addressing the needs for expanded frequency spectra for wireless communications services there are essentially only a limited number of options that can be pursued. These are to follow one or more of the following strategies: (a) allocate new frequency bands for new services, especially in the largely unexploited millimeter wave band; (b) utilize the unregulated infrared or optical frequencies; (c) reallocate spectra over a transitional period from users who are using frequencies ineffectively to higher-value users who would use these resources more intensively (this is sometimes called the "use it or lose it rule"); (d) move certain band allocations from single-use to multiuse rules with new frequency sharing criteria; and (d) create new rules and guidelines for more intensive and effective use of the spectra now in use, particularly with regard to multi-use/multipurpose allocations.

With new digital encoding techniques such as CDMA it may very well be possible to initiate more terrestrial and space communications systems of frequency sharing. Even in the area of space communications, with some 17 defined space services it may be possible to allow a number of different satellite applications to share the same band. The current major bands allocated to space communications now operationally available are provided in Table 6.3. Even allowing for the fact that there are additional new allocations still to be implemented from WARC 1992 in the coming years, two key conclusions seem obvious even now. First, there is indeed a great deal of spectra covering many, many gigahertz of bandwidth already allocated to space communications service. Second, these allocations may very well be excessively segmented in terms of restricting the maximum flexible use of this limited resource. Although these observations are among the most obvious other broadened definitions and greater reliance on multipurpose frequency allocations should likely be considered in other areas as well.[13]

New "efficiency" guidelines might well include digital compression techniques, higher-efficiency modulation, encoding and multiplexing techniques, cellular frequency reuse techniques, polarization discrimination reuse techniques, and physical separation of services. The opportunities for physical separation between different types of services are considerable. One might well be able to use the above techniques so as to reuse frequencies for geosynchronous, medium and low earth orbit satellites, proto-space users (i.e., High Altitude Long Endurance (HALE) platforms), and terrestrial services.

It could even ultimately lead to a new system of noninterfering frequency use that would eliminate the need to allocate frequencies at all. While this may become technically feasible in 20 to 30 years, the importance of national "political" control of spectra and the economic and financial dimensions of frequencies as a revenue generator cannot and will not be easily discarded. This idea of self-assigned frequencies on a noninterfering basis is thus technically appealing, but probably impossible from a policy perspective for some years to come. In short this is a very innovative concept that will take many years to negotiate and implement if it becomes possible at all. Further, if it were implemented at the in-

Space-to-earth (down-link)	Earth-to-space (up-link)
Fixed satellite service	
2500–2655 MHz (Region 2) 2500–2535 MHz (Region 3) 2655–2690 MHz (Region 2, Region 3) 3400–4200 MHz 4500–4800 MHz 7250–7750 MHz 10.7–11.7 GHz 12.5–12.75 GHz (Region 1, Region 3) 11.7–12.3 GHz (Region 2) 17.7–21.2 GHz	5725–5850 MHz (Region 1) 5850–7075 MHz 7900–8400 MHz 12.75–13.25 GHz 14.0–14.8 GHz 17.3–18.1 GHz 27.0–27.5 GHz (Region 2, Region 3) 27.5–31.0 GHz
Maritime mobile satellite	
1530–1544 MHz	1626.5–1645 MHz
Aeronautical mobile satellite	
1545–1559 MHz	1646.5–1660.5 MHz
Mobile satellite	
1544–1545 MHz 19.7–21.2 GHz	1645.5–1646.5 MHz 29.5–31.0 GHz
Broadcasting satellite	
2500–2690 MHz 11.7–12.5 GHz (Region 1) 12.1–12.7 GHz (Region 2) 11.7–12.2 GHz (Region 3) 22.5–23.0 GHz (Region 2, Region 3)	

TABLE 6.3　FREQUENCY ALLOCATIONS FOR SATELLITE COMMUNICATIONS

ternational level there is still the national regulatory processes to consider and vice versa.[14]

It should be noted that all of these techniques can work relatively efficiently when all of the users of the frequencies are under the control of a single regulatory authority, such as the FCC in the United States or OFTEL in the United Kingdom. Problems can emerge when you have a number of adjacent countries attempting to regulate and control frequency use according to different standards and allocation schemes. The International Telecommunication Union is not well equipped to "enforce" the international rules and recommendations it helps to develop since there are no effective sanctions in place. It is possible that at some point in the future trade agreements will include provisions for enforcement of international frequency allocations.[15]

Even in the United States this can be a problem with regard to the adjacent countries of Mexico, Canada, and a number of Caribbean countries. As a practical matter, the United States and Canada have had difficulty reaching an agreement with regard to the new allocation of frequencies above 2 GHz for the new DBS radio broadcasting service. Since this is intended to be a national service with a satellite footprint that will cover all of the United States and by spill over some 80 percent of the population of Canada along the U.S. border line this becomes a serious international regulatory matter.[16]

Specific illustrations of how each of these strategies are currently being implemented are provided below, as well as an assessment of the longer-term effectiveness of these strategies. Many experts believe that the most technically demanding of these options, namely the development of the new upper frequency bands in the millimeter wave band, will see Japan and Europe leading the United States in this area. In a survey of U.S. industry officials in the field of telecommunications, American-based R&D to exploit frequencies in the range of 30–300 GHz was seen as lagging.

In particular, officials of some 40 U.S. telecommunications organizations, when recently interviewed on this subject, expressed the views that Japanese and then European technology would lead the way in these areas. Given the rapid strides being made in millimeter wave technology research with the COMETS programs in Japan and the ITALSAT project in Europe, and the lack of an active research program in the United States by NASA or others, this seems to be a very possible projection for the future.[17]

A more detailed review of the four prime strategies for addressing the frequency shortage is provided below. It should be noted at the outset that a combination of these approaches is more likely than just one single approach being adopted.

6.3.1 Allocate New Frequency Bands for New Services

The idea of allocating new frequency bands for new service sounds like a very straightforward and easy idea to implement. The problem is that there are not many frequencies to allocate. The alternatives are to designate the frequencies in the millimeter wave area for newly established service categories or allow a new type of user category to somehow share, overlap, or fit in between existing allocations.

The FCC allocation of frequencies for tests and demonstrations in cellular television in the millimeter wave band is an example of a new allocation based upon a new technology and expanding demand for wireless television. The FCC's allocation of frequencies near 2 GHz for both ground and space mobile communications on a shared use and subsequent negotiations to agree how to achieve noninterference is an example of a new "shared use" allocation within a previously exclusively allocated band. The development of new technology can allow more intensive sharing of frequency bands to occur and may lead to more multiple-use applications in the future. The lack of precision and seemingly nonstructured approach is a departure from past practice, but the advances in technology suggest that "operationally defined sharing" techniques can indeed be successful.

Such sharing is, in fact, facilitated by a number of technologies which, as noted above, may in the future even be specified in either recommended or even mandatory guidelines. These technologies today include better directionality in antennas, higher-performance digital modulation and encoding techniques, and the application of advanced coding and processing techniques. The future direction in frequency allocation will thus likely see more and more sharing among terrestrial telecommunications systems; HALE platforms; LEO, MEO, and GEO satellites. This trend toward sharing will also see more and more hybrid and integrated services. These will likely include cellular voice, data, paging, and GPS navigational services provided through a single terminal. Geographical delimited multipurpose uses may also evolve in countries that have imposed more strict technical guidelines for the sharing of frequencies.[18]

The biggest single problem with such new frequency allocations is likely to be a lack of international coordination in how the spectrum is assigned. In the case of millimeter wave allocations for cellular television the United Kingdom is planning to designate a band at 44 GHz, while the United States is debating a 2-GHz-wide band starting at 28.5 GHz. Tests of such cellular television have proven the feasibility of such technology and thus there is strong pressure to decide this matter in the near-term future. The orderly process whereby the ITU World Administrative Radio Conference sits down and agrees on a global allocation for new service seems all but remote history in today's rather turbulent times.[19]

The tendency today is more toward regional allocations (in Regions 1,2, and 3), and even then the ITU Radio Regulations are riddled with footnotes where exception after exception is taken. The reasons for such footnote exceptions some 20 years ago were very limited and involved such reasons as key national defense applications. Today, exceptions can and are taken for almost any reason. Desires to avoid equipment retrofits, plans to broadcast in adjacent frequencies to those already in use, even desires to carry out surveillance on other countries have been the overt or covert reasons for exceptions being taken.

Even if exceptions are not taken, a country still has a number of options available to it. All uses of spectra are available for exploitation by nations under the ITU procedures on a so-called noninterference basis. Further, experimental and trial use licenses can be issued by governments when a new or reallocated use of spectra is being investigated. For years, in the former Soviet Union frequencies were used for combined military, governmental, and commercial purposes, often with common systems and frequencies. These programs, including satellite launches, were conducted without any ITU filing or notification. In the vast regions of the Soviet Union, the national government thus became their own ITU.[20]

This process of ITU "recommendations" in lieu of mandatory rules could well become increasingly troublesome in the twenty-first century. The concept of countries or even private corporations proceeding with the implementation of wireless service in a nondesignated band on the basis of a test or experimental basis and just the presumption of noninterference, is an interesting but potentially dangerous historical artifact. For many years the ITU was almost like a "club," where members were on their honors to do the right thing. In the old environment only a few countries were involved and recommendations once made were very broadly observed. Things have changed.[21]

Today telecommunications is headed toward becoming a trillion dollar industry and stakes are increasingly high. The scramble for more and more frequency allocations and the lack of effective sanctions creates a real dilemma. Enforcement procedures and mandatory registry of all uses whether formal, test and demonstration, or "noninterference" applications, are clearly needed. The ITU is essentially a technical organization with a long tradition in this area is clearly underequipped for imposing sanctions for violators of the international frequency allocation and usage procedures.[22]

This is not a small problem. It has been seriously suggested that the ITU as a technical coordination body and a standards-making group is not well equipped for an international enforcement role, particularly in an environment characterized by more and more competition. The issue is more closely related to equitable and enforceable trade agreements than it is to technical standards. One proposal is that the new World Trade Organization, which will replace the General Agreement on Trade and Tariffs (GATT), should be given the responsibility of enforcing international violations of frequency allocations.[23]

6.3.2 Reallocation of Spectra

Over a transitional period, spectra may very well be reallocated from users who are using frequencies ineffectively to those who will make more effective use of the spectra. Exactly what is more efficient use of spectrum may be difficult to objectively determine, but at least one can assume the allocation will be to higher value users. There are certainly some objective measures, such as information rate per hertz of capacity. This could measure the efficiency of the modulation and encoding techniques and other aspects of the transmission and switching technology. Ultimately the issue is more likely to be related to who will pay the most.

The more difficult issue is the relative value or merit of different applications in relation to the size of the market they service. Is a television channel showing cartoons to 3 million children a higher-value user than an Instructional Fixed Television Service channel that is teaching physics to 5000 high school students? Is a lawyer talking to a client on a cellular phone a lower-value service than a 911 call from an injured car accident victim in a remote area? Any attempt to quantitatively compare commercial telecommunications services with socially based, scientific, defense, or humanitarian applications is almost guaranteed to end up in a hopeless morass. As long as technology allows more and more efficient and expanded use of existing and new frequency bands, it seems likely that balanced and "reasonable" accommodation of all types of applications will continue to prevail as a pragmatic rule of thumb.

The most likely result is that some services that involve national defense, public safety, or other critical social services will remain with a high-value status that will be little challenged. Nonutilized or underutilized frequency bands in the military sector may be the exception to the rule. The remainder of the commercial services, however, could well be put into economic competition with one another. The FCC might very well identify lower levels of usage and lower-value bands of spectra where it proposed a migration from one band to another over a period of time. This would allow lower-value users to relinquish certain bands that could then be auctioned off to higher-value users. If recent past experience is indicative of how this might be done, the revenues realized from the auction can and indeed will be used to assist the migration of the previous users to new bands. Again, pragmatism and mutual accommodation rather than highly theoretical or deterministic formulas seem the most likely to proceed.[24]

In some countries, the auction process may be judged inappropriate, but nevertheless there may well be a need for some frequency reallocations. This may be especially true for wireless mobile services. Here the issue may be in deciding what frequencies are being poorly used on a combined technical, operational, social, and business basis and which new applications are in need of receiving new frequency bands. Most countries operate within the ITU Radio Regulations and observe the international frequency allocation tables. Given this broad level of international agreement as to what frequencies should be used for what purposes, there are often limited choices of what "innovative or unconventional" reallocations of frequencies can be "legally" carried out through this process.[25]

One of the biggest costs of conversions from one frequency band to another is the need to change the frequencies of transceivers or simple receivers. If the service should involve a mass or population-wide application such as radio or television, then millions of units can be involved. In such cases the transfer of frequencies from one band to another can vary anywhere from very difficult to virtually impossible.

There is, looking to the future, another approach that would allow much greater flexibility. This is the concept sometimes called the software-defined radio. This would have a digital processor define the frequencies utilized by a particular unit. This same processor could be instructed to retune the operating frequencies and thus easily migrate to another band or even to allow a different type of service. Such an appealing technical solution is still not easily implemented. Even if the shift were made beginning almost at once, it could easily be ten, fifteen, or even more years, before these new highly flexible products were in sufficient widespread use that the switch-over could be made.[26]

6.3.3 Multiuse Frequency Bands

The insatiable demand for more frequencies will probably also lead to the conversion of certain band allocations from single-use to multiuse rules with new frequency-sharing criteria. For years the allocation of frequencies was relegated to the domain of engineers who largely worked for monopolized telecommunications organizations. This somewhat isolated and protected environment let standards become quite cautious and stable. The use of conservative guard bands, wide channelization, and sound but limited efficiency transmission systems were reflective of the practices agreed to and provided by AT&T, the Bundespost, the British Post Office, and so on.

In the mid-1980s this environment changed, and changed rapidly. This new period of divestiture, deregulation, privatization, and competitive telecommunications services has altered the frequency allocations process. For instance, the allocations of frequencies at 2 GHz for LEO and MEO mobile satellite services and for PCS would probably not have happened in a noncompetitive environment. This new business environment as well as new technology are driving the world of spectra management in new directions.[27]

This idea, sometimes described as multiservice or multipurpose allocations, suggests changing the approaches of the past. These changes would allow different users to take advantage of new digital technologies, phased array antenna technologies, and new processing capabilities. This new philosophy is that narrow, precise, and conservative definition of a service with an exclusive allocation may limit the use of a frequency band in unnecessary ways.

The advent of the CDMA spread-spectrum-type multiplexing systems can allow operation in a very-high-noise urban environment while also involving low levels of interference to others. The idea is to find a wider range of wireless telecommunications services on the ground, in the air, and in space that can coexist. In essence the factors that are most important are power levels, patterns of frequency use and reuse, coding and processing techniques, and geographic separation or limited overlap. As noted earlier in this chapter, the possibility of a broadened definition for multiuse/multipurpose space communications allocations and greater sharing of terrestrial and space services could both be productive areas to study in this regard.[28]

6.3.4 Creation of New Rules and Guidelines for More Intensive and Effective Use of the Spectra Now in Use

The world of spectrum allocation has gone from an obscure and not very significant enterprise to a multibillion dollar activity. Decisions on who gets what spectra can virtually make or break a wireless telecommunications enterprise. The development of new and better procedures for the allocation, shared multiple use, reassignment, and transfer of the user rights of spectra would likely help the overall development of the wireless industry.

The challenge in this regard is to obtain the following results: (a) create the correct incentives or penalties to encourage efficient and expanded use of the spectrum, (b) allow more sharing, and (c) move inefficient users into different and potentially smaller bands. It is a further challenge to do the above without disturbing vital public services, creating inequities among different classes and types of users, or by mortgaging the future by denying opportunities for future expansion.

The key to achieving this type of strategic plan for the future is most likely to be financial. Rewards for the use of the most effective technology are clearly effective. The renewal or award of licenses can also be given on the basis of frequency efficiency, upgrades in digital equipment, or enhanced patterns of frequency reuse. The "carrot" can also be supplemented with the "stick." Owners and operators of radio and television stations may, for instance, be refused renewal of their license if they do not retrofit their equipment to achieve higher levels of efficiency over a reasonable time of transition. The reduction of frequency inefficiencies coupled with aggressive development of new higher frequencies will probably both be needed to meet the challenges of the twenty-first century.

The disparity of practices among various nations and the lack of control of the International Telecommunications Union in the area of frequency allocation and use will create major problems as wireless services continue to grow in coming years. The issues of Tongasat and Pacstar-like surrogates (e.g., Papua New Guinea) in registering frequencies will certainly be a challenge that could easily pit developed and developing countries against each other in future years. Likewise the internationalization of the frequency auctioning process could be an even larger wedge between those with economic resources and those without.[29]

The irony is that wireless technology in terms of satellite and wireless cellular technology is in many ways ideally adapted to application in the urban and rural areas of developing countries. Joint ventures involving wireless enterprises from the OECD countries together with developing countries may well serve as a mechanism for introducing this technology more rapidly around the globe and to encourage frequency efficiencies as well. The desire for flexible, mobile, low-cost, and broad band services can likely only be achieved through a systematic attempt to increase spectrum efficiency through much higher throughput per hertz, combined with new allocations. Efficiency guidelines for frequency use that are internationally adopted could serve as a key stimulus to progress in this area.[30]

6.4 CONCLUSIONS

The electromagnetic spectrum is a rare commodity of increasing value, which has the remarkable ability to renew itself. The more one uses this resource the more valuable it becomes. Today as the demand for wireless services grows rapidly, the technology is allowing many ways to expand the derived capacity. These innovations include digital compression; intensive modes of frequency reuse; enhanced modulation, encoding, and multiplexing techniques; and complex patterns of sharing. Adding additional interest to this process is the move to use spectrum auction as a method of transferring outmoded allocations to newer and more valuable applications.

In the future, spectrum use seems likely to become more and more complex. This means international coordination will be more important. It means sharing among intraoffice, terrestrial cellular, HALE-based platform services, and satellites in LEO, MEO, and GEO will become more and more prevalent and multipurpose allocations will also be used for more and more services and applications.

As the entire field of wireless telecommunications moves upward of $200 billion per year in total global sales, the value of spectra as a global resource could be considered to reach staggering levels. The national and international mechanism created to address spectra matters were implemented in a time of simpler technology, of agreed-upon principles of international collaboration, and telecommunications monopolies. These mechanisms are not necessarily well adapted to today's increasingly competitive telecommunications environment. In particular, the ITU does not seem well suited to an active enforcement role with regard to "illegal" or improper use of spectra. In the future, international trade regulation and controls may indeed be needed to enforce effective and consistent patterns of spectra use.

ENDNOTES

(1) Freeman, R.L., *Telecommunications System Engineering* (New York: Wiley, 1981), p. 20ff.

(2) ———, *Telecommunications System Engineering* (New York: Wiley, 1981), pp. 30-35.

(3) Bellamy, J.C., *Digital Telephony* (New York: Wiley, 1982).

(4) "Fundamental of Radio Communications" (U.S. Department of Commerce, Institute for Telecommunications Science, Boulder, CO: U.S. Government Printing Office, 1990), pp. 20-24.

(5) "Fundamental of Radio Communications" (U.S. Department of Commerce, Institute for Telecommunications Science, Boulder, CO: U.S. Government Printing Office, 1990), pp. 26-27.

(6) "Fundamental of Radio Communications" (U.S. Department of Commerce, Institute for Telecommunications Science, Boulder, CO: U.S. Government Printing Office, 1990), pp. 28-30.

(7) Taylor, L., "Radio Frequency Auction Risks Future of Satellite Communications Allocations," *Signals*, September 1993, pp. 1-2.

(8) "PCS Bidding Rules Issued: Messaging Services To Be Licensed This Summer," *Inside Wireless*, April 27, 1994, Vol. 2, Issue 7, pp. 1-2.

(9) "WARC 1995," *Telecommunications Journal* (Geneva, Switzerland: ITU, March, 1994).

(10) "The Radio Frequency Spectrum (ITU Region 2)" (Denver, CO: Communications Technology Inc., 1993), pp. 1-2.

(11) "The Radio Frequency Spectrum (ITU Region 2)" (Denver, CO: Communications Technology Inc., 1993), pp. 3-4.

(12) "The Case for Multi-use Spectrum Allocation: The Problem of Excessive Segmentation in Frequency Allocations" (Washington, DC: NASA Office of Advanced Concepts and Technology, 1992), pp. 1-8.

(13) Taylor, L., "Radio Frequency Auction Risks Future of Satellite Communications Allocations," *Signals*, September 1993, pp. 1-2.

(14) Calhoun, G., *Modern Cellular Radio* (Norwood, MA: Artech Publishing, 1987), pp. 109-120.

(15) *The Office of Telecommunications (OFTEL) Annual Report* (London, U.K.: Department of Industries, 1993).

(16) "US Position on Radio Direct Broadcasting Services in the 2 GHz Band," FCC Filing, Wiley, Rein, and Fielding, April 1994.

(17) "Planning and Research for Millimeter Wave Application in Satellite Communications: U. S. Industry Survey," University of Colorado, Center for Advanced Research in Telecommunications, May 1993 (unpublished research report).

(18) "Multi-Purpose and Shared-Use Frequency Allocations," *NASA Technical Report* (Washington, DC: NASA, 1992), pp. 40-52.

(19) Codding, G., "ITU Radio Regulations, WARC and the Frequency Allocation Process" (London, U.K.: International Institute of Communications, 1992), pp. 2-4.

(20) ——, "ITU Radio Regulations, WARC and the Frequency Allocation Process" (London, U.K.: International Institute of Communications, 1992), pp. 5-6.

(21) ——, "ITU Radio Regulations, WARC and the Frequency Allocation Process" (London, U.K.: International Institute of Communications, 1992), p. 7.

(22) ——, "ITU Radio Regulations, WARC and the Frequency Allocation Process" (London, U.K.: International Institute of Communications, 1992), pp. 9-10.

(23) Pelton, J.N., "Comparative Overview of Satellite Communications: Trends in Regulation and Competition," 45th Congress International Astronautical Federation, Jerusalem, Israel, October 1994, pp. 1-8.

(24) Vion, R., "Improving the Efficiency of RF Frequency Use," *IEEE Communications Magazine*, June 1993, pp. 58-65.

(25) ——, "Improving the Efficiency of RF Frequency Use," *IEEE Communications Magazine*, June 1993, pp. 66-68.

(26) Gilder, G., "Software Defined Radio," Executive Telecommunications Forum, Seattle, WA, January 1994.

(27) "Multi-Purpose and Shared-Use Frequency Allocations," *NASA Technical Report* (Washington, DC: NASA, 1992), pp. 53-56; see also Gilder, G., "Competition Creates More Efficient Use of Spectrum," *The Economist*, September 15, 1991, pp. 27-34.

(28) Parker, E., "The Technical and Operational Advantages of CDMA Encoding Techniques" (Washington, DC, Phillips Publishing, 1991), pp. 12-13.

(29) Taylor, L., "Radio Frequency Auction Risks Future of Satellite Communications Allocations," *Signals*, September 1993, pp. 1-3.

(30) Pelton, J.N., "Comparative Overview of Satellite Communications: Trends in Regulation and Competition," 45th Congress International Astronautical Federation, Jerusalem, Israel, October 1994; also see "The Need for Stricter Frequency Efficiency Rules and Sanctions for Violations of the International Frequency Allocation Process," *Via Satellite*, Editorial, August 1994.

CHAPTER SEVEN

POLICY, REGULATORY, AND MANAGEMENT ISSUES IN WIRELESS SERVICES

7.1 INTRODUCTION

Only two decades ago mobile terrestrial wireless services represented an almost negligible amount of the world's total commercial telecommunications traffic. This was a service for dispatching taxicabs, maintaining fleet communications for trucking firms and service vehicles, and for a limited amount of VIP mobile communications for limousines. The rest was essentially military communications or mobile links for police, fire, or emergency support. Only a very few individuals had mobile radio telephone service. This was in a large part because there were few frequencies available for this purpose, and those that were available were very inefficiently used. The creation of cellular radio telephone service in the early 1980s clearly changed all this. In fact, it has given birth to a remarkable new multibillion dollar market in the process.

Today, frequencies for future PCS mobile services have been auctioned by the U.S. government for billions of dollars. New wireless services are cropping up everywhere. In the process the competitive model of regulation is being given its fullest test. The new PCS service will have not only multiple standards but a much larger number of competitors in each market area than ever before. It will test just how much competition is optimum for the consuming society.

There are also numerous proposals for low earth orbit mobile satellite systems and a growing number of Direct Broadcast Satellite systems around the world. There are new cellular radio services at the millimeter wave band that are being opened up for broadcast and mobile communications with experimental trials currently in process. Specialized Mobile Radio services have gone from being a minor league "mom and pop" operation to a major multibillion dollar business with organizations such as NEXTEL planning national coverage. There is now more competition and more competitive services in the wireless sector than ever before in the history of telecommunications.

The turmoil of the marketplace, the swift pace of the technology, and especially, the rapid vicissitudes of the policy and regulatory framework for wireless communications in the United States and abroad are extremely difficult to follow. The total environment, especially in the policy and regulatory arena, is in a word volatile. Over $20 billion was invested in cellular radio during the last eight years in the United States and it is likely that over five times that amount will be invested in various forms of wireless communications within the next decade. This is driven in part by continuing market demand, but it will also be driven by the huge amounts of money that were invested to acquire PCS spectra through the recent auctioning process as well. In short, huge investments in the acquisition of new frequencies as well as relocation costs will mandate proportionate investment in the attempt to generate new revenues as soon as possible.[1]

This chapter reviews the latest changes and innovations in the regulatory framework as well as plots the more important trend lines for the future. Although there are a number of key standards issues that impinge on policy and regulatory sectors, these are separately addressed in Chapter 8. Here the focus is on the legislative, judicial, and regulatory issues for wireless telecommunications in the United States and to a lesser extent in Europe and the Asia-Pacific region. It also addresses innovations in modern management techniques currently being applied to the telecommunications industry. Management and policy issues are often closely linked and in the modern competitive environment, the need to adopt modern management techniques that are consumer responsive is crucial to success. In particular, the most important management innovations that are directly tied to competitive markets are reviewed. In short, just as regulatory policy is shifting toward marketplace- and customer-led control mechanisms, the latest management trends in wireless telecommunications appear to be headed in the same direction.

Initially the prime focus will be on the key policy and regulatory issues as outlined below. This will be followed by a complementary analysis of the key management issues. These issues include:

(a) current policy and regulatory control, especially with regard to tariffing and charging policy for wireless service offerings both for common carriers and private network providers;[2]

(b) institutional arrangements for the provision of wireless services covering both direct providers of networks and services as well as value-added carriers;

(c) specific regulatory authorizations, especially with regard to obtaining and retaining construction permits and licenses for the provision of wireless service;[3]

(d) the increasingly popular unlicensed wireless industry;

(e) wireless service issues that impact on social issues such as "911" performance and coverage, mobile-based services for the handicapped, or even electronic crime that might be facilitated or prevented by mobile services; and

(f) important new U.S. legislation concerning the regulatory liberalization of the telecommunications sector, especially from the perspective of the wireless communications industry. (This primarily will address proposed new legislation that would revise key elements of the Modified Final Judgment of Judge Harold Greene.)

Finally, this chapter will conclude with a brief examination of the world regulatory environment that will include a comparative look at how U.S. practices are in contrast or are similar to the rest of the world.

7.2 KEY POLICY ISSUES AND REGULATORY CONTROLS

This section covers, among other subjects, financial matters, procedural approvals, licensing, frequency allocation, public policy issues, health-related standards, and competitive practices vis-à-vis other countries.

7.2.1 Important Tariffing and Charging Policies for Common Carriers and Private Network Providers

The world of wireless telecommunications, when it comes to constructing "logical" tariffing structures, is somewhat puzzling to those who have been previously concerned with wire and cable networks that are linked to hierarchically switched services. In the traditional wire-based world, when you install more switches and more wire it almost always produces two clear results. First, one can provide more and probably better service as a result of the expanded capacity. Second, a specific amount of money has been spent to achieve this increased capability. This means that the installation of longer and higher-capacity links as well as more switches also costs more and more money and in a fairly directly linear fashion—double the capacity or double the distance covered and you tend to double the plant costs. It follows in such a linear environment that increased charges will be made to cover these increased costs.

The wireless world, which is dependent upon radio transmission, however, is not nearly as cause-and-effect oriented. If you create a satellite system with, say, global beam coverage, it is essentially the same amount of satellite capacity that is used regardless of distance—at least up to 40 percent around the globe. If one calls from Cleveland, Ohio to Toledo, Ohio, the same channel and power is utilized as if the call were to go from Cleveland, Ohio to Toledo, Spain or even Kinshasa, Zaire. In the case of a cellular telephone network, the rules of investment costs are a little more orderly than in the rather extreme case of a satellite system. This is because each new cell tends to cost a fixed amount more. The costs of a cell operation are relatively fixed. If there is one car operating in a cell or 500 automobiles operating in it, the capital and operating expenses are largely the same. Coping with busy hours of operation and cell saturation are very critical to a successful mobile telecommunications operation and this is why premium rates for busy hours are imposed.[4]

Once a wireless telecommunications is established for a fixed-service area, the basic network costs are fairly stable. The distance covered within the "radio footprint" does not affect the costs. Likewise, the number of signals being served, at least up to the system's limits, does not affect the costs either. Once saturation is reached, however, this is clearly a key service problem. Higher peak period rates are one common response. The technical solution involves cell splitting of congested zones. This is the most common strategy for dealing with this problem. In many ways wire-based systems are distance-sensitive and investment plant sensitive in ways that wireless systems are not.[5]

Wireless systems with five to eight mile cells are in many ways distance insensitive. The new PCS or microcellular systems of the future will tend to reverse this trend, but even here the economics are much different from wire-based systems both in terms of build-out and operational costs. Further, wireless systems tend to be different from wire-based systems in their ability to be redeployed. The available capacity in a wireless system can usually be redistributed among many different possible pathways in millisecond increments to reflect changing demands. Wire-based systems that provide fixed "pipes" to fixed locations are much more static.

In very simple terms, the cost structures for wire-based terrestrial systems and wireless-based radio systems are for technical and physical construction reasons much different. Furthermore, the sense of physical ownership and control is higher with the "tangible hardware" aspects of wire- and cable-based systems. In contrast, the more "intangible" wireless systems just radiate through free space and thus seem less "real." Frequency allocations are often made available to wireless carriers for only certain periods of time and may be reassigned to other uses or carriers. Again, this serves to create a feeling of less ownership and control.[6]

Finally, wireless systems are often seen as a "bypass" technology. They are often perceived by established wire carriers to be an enemy of established terrestrial telephone networks. The acquisition of McCaw Communications by AT&T to obtain direct access to customers without using RBOC networks is a clear case in point. The license to send a signal through the air or space does not seem to be the same as "ownership." It certainly cannot be "protected" as an asset in the same way that a wire, or a cable, or a fiber system can. These wire-based networks are much more tangible. They are not only "real," but are also physically linked to a network of switches.[7]

For the above reasons, it is not surprising that the charging concepts for wireless systems tend to be different from the more traditional services. In particular, most subscribers to cellular radio pay for equipment they require for their personal needs, plus a fixed monthly fee to access the service, and then a per-minute charge for actual usage. This per-minute rate often varies according to patterns and volumes of use during the day and night. The subscribers also normally pay for all *incoming and outgoing* calls. This tends to make cellular subscribers only inclined to give out their number to close friends and business partners, and to get rather angry when called by telephone salespeople. As cellular, personal communications, and advanced paging systems become more widespread and as the trend toward universal personal numbers become more pervasive, these charging concepts will likely evolve and perhaps will be simplified. From the opposite perspective, however, conventional services may move to resemble cellular rate structures. This can be seen in the move toward pay-per-view on cable television and in efforts by telephone companies to move away from basic flat rates and toward per-minute charges on the basis of universal measured service charges. Most state public utility commissions have been reluctant to authorize such rates.[8]

Today, in the U.S. public switched network, or what is now officially called "commercial mobile radio service" (CMRS) providers must file their tariffs with the FCC. In contrast, those providers deemed to be "private mobile radio service" (PMRS) providers are not regulated in this way and do not have to file tariffs. The chart shown in Figure 7.1 indicates how the FCC has divided the two types of CMRS (commercial) and PMRS (private) services.

▶ Commercial Mobile Radio Services ("CMRS")

   • For profit AND interconnected AND available to the public

      OR

   • Functionally equivalent to CMRS

▶ Private Mobile Radio Services ("PMRS")

   • All other mobile services

FIGURE 7.1   Commercial and Private Mobile Radio Service
                (CMRS and PMRS)

The primary mechanism to keep cellular tariffs reasonable over the past few years has been the concept adopted by the FCC of awarding two licenses or local cellular frequencies so as to achieve competition in each major urban market.[9] With the advent of the new Personal Communications Service, the FCC in response to a U.S. Congressional mandate, has proceeded to auction frequencies for this service to an increasing number of carriers. This process will ultimately see cellular service providers, specialized mobile radio service providers, PCS systems, and to a certain extent, low earth orbit mobile satellite networks all essentially competing for mobile services. Presumably this robust competition will serve to lower costs and stimulate new services.

There will also likely be increased mobile services, especially for data communications in the unlicensed bands. Some economists and market analysts have suggested that there is not sufficient consumer and business communications demand to support this many competitors for mobile services. The FCC has, however, concluded that they would risk some oversupply in the market and even some bankruptcies rather than have too little competition.[10]

One of the key regulatory issues in terms of tariffing arrangements is that the cost of accessing frequencies will be different for localized as opposed to regional or nationally based mobile services. Apparently the decisions by Motorola to invest heavily in the NEX-TEL Specialized Mobile Radio Service was in large part driven by the lower cost of obtaining frequencies in the SMR bands for a nationally based service as opposed to the frequency auction method. Further, the decision by some vendors to enter the unlicensed bands rather than enter the PCS market was also apparently strongly influenced by this factor as well.[11]

Today, the trends in business communications are still moving toward the creation of private networks. This means not only office-based Local Area Networks (LANs), which may be wire or wireless in their design, but increasingly, Wide Area Networks, or so-called Enterprise networks for linking even global corporations. The most extensive of these networks can even be called Global Area Networks (GANs).[12] Recently a new International Computer Telephony Association has been formed to champion the cause of private, personal computer-based telecommunications networks.[13] This all suggests that in the future there may be more and more private-based wireless networks and that the software aspects of such networks may be the most important commercial dimension. These may, in fact, be wireless or wire-based PABX or LAN systems located on various corporate campuses that are in turn linked together through Wide Area Networks (WANs) to national and even global locations. It is possible that an individual will have a single hand-held device that operates within the corporate private network, but that can automatically seek and switch to a publicly tariffed PCS, SMR, or cellular service provider when the private network is not available. Alternatively, one might also be able to access unlicensed wireless wide area networks (WANs), which one or more vendors may establish in many major urban markets.

These private/public interface arrangements for future wireless service, in terms of the technical architecture and charging arrangements, are thus likely to become increasingly complex. The ability of private networks to offer a richer mix of customized services and lower-cost offerings based upon corporate economies of scale will exert great pressure on the public mobile service providers to reduce their tariffs to the lowest possible price. This suggests that competition in public mobile telecommunications services will certainly serve to drive down costs and tariffs for these services, but the private networks may even have a greater impact. This same phenomena in terms of public and private mobile networks will likely be seen in the area of satellite mobile services as well.[14]

7.2.2 Institutional Arrangements for Network Providers and Value-Added Carriers

There are essentially five different institutional frameworks within which wireless services can be provided. These are: (a) a government or military agency; (b) international agencies or corporations with the authorization to operate a domestic, regional, or global service or their designated local entity; (c) an approved and licensed public telecommunications carrier; (d) a licensed, private network carrier or value-added carrier; or (e) an unlicensed operator of network service who is utilizing frequencies that have been designated as open-entry noninterference bands.

Examples of each type of wireless telecommunications operator would include the following:

(a) Government or military agencies: the Department of Defense Milstar System; the Office of Emergency Preparedness National Emergency Broadcasting System, the National Forestry Communications Network, and the United States Information Service (e.g., Worldnet);

(b) International agencies or corporations: the International Amateur Radio Association's Oscar Satellite System, INTELSAT, INMARSAT, British Sky Broadcasting, Asia Sat, EUTELSAT, European Broadcasting Union, NATO Skynet System, PanAmSat, Orion, Globalstar, Teledesic, and Iridium;

(c) Telecommunications and Broadcasting Carriers: US WEST Cellular, Cellular One, Public Broadcasting System, National Public Radio, MCI, ABC, TBS, Sprint, GTE, ARINC, and so on;

(d) Private, Licensed Wireless Networks: the National Technological University, Europace, the New York Teleport, WAL-MART, General Motors Corporation, Holiday Inn, and so on;

(e) Unlicensed private networks: CB radio operators, unlicensed wireless data services in the PCS band, TVRO owners, corporate headquarters equipped with wireless PABX systems (e.g., systems provided by Spectrallink or similar suppliers), walkie-talkies, and so on.

Today the great regulatory emphasis is placed on the public telecommunications carriers. This is largely because governmental agencies and military units as well as international agencies have fairly well-defined self-control procedures in effect. Further, the unlicensed service is structured to be largely self disciplinary and the private networks have in the past been a small part of the total market. There are, however, some flaws in the regulatory scheme. First, governmental and military users have tended to retain large chunks of the spectrum, which is often sparsely and ineffectively used. Recently, there have been attempts to try to reallocate some of this underutilized private and public spectra and seek more intensive usage patterns and to derive higher value thereby.[15]

A second flaw is in the rapid growth of private networks. As private networks with seamless interconnection to the PSTN continue to grow and expand, there will be more regulatory focus on these telecommunications markets. These regulations will cover localized customer-premise wireless equipment all the way on up to the new "private international satellite systems," such as PanAmSat, Orion, Iridium, Globalstar, Teledesic, and so on. All of these private networks and more will become the focus of new regulatory concerns. These concerns will include interconnection protocols, the extent and nature of any cross-subsidies that might exist across the CMRS/PMRS interface, and equality of regulatory treatment across public/private boundaries in the wireless telecommunications environment.[16]

These private networks also pose questions concerning major issues of frequency assignments and orbital assignments. Other concerns include health standards, transborder data flow, "open versus proprietary" interfaces, technical and quality standards, discriminatory pricing and tariffing concepts, international trade issues, and regulatory restrictions and priorities for primary network providers versus value-added carriers. There is concern, at least in some countries, that private networks as promoted by such groups as the International Telecommunications User Group (INTUG) is simply a means of bypassing the standards and regulatory controls that have been set for public telecommunications carriers. In fact, private networks can provide bypass at the local, the national, the regional or the global level. At the local level this can be accomplished via wireless LANs or it can be a fiber-based loop in the form of Alternative Network Providers (ANPs) or Competitive Access Providers (CAPs). At the national or international level the wireless "bypass" is more likely to be a satellite network.

It is possible, at least in theory, for an integrated bypass system to use a combination of wireless LANs, cellular telephone systems, CAP- or ANP-based fiber loops, satellite teleports, and private satellite networks to bypass public telecommunications carriers at any level of a telecommunications network.[17]

One could take the example of a small company that installs a wireless PABX/LAN network in their offices. This company could sign up to interconnect with a private business fiber loop that connects to a regional teleport. This small business could then route all of its overseas telephone, fax, videoconference, multimedia, and data traffic via, say, the PanAm-Sat private satellite network . This hookup could then reach locations in South and Central America and Europe. If one were a much larger firm, the entire enterprise network that links together all corporate offices and industrial plants could be either entirely internally owned and operated or leased as a virtual private network. These private networks may use wireless LANs and PABXs, support wireless mobile data networks in the licensed bands, and interconnect via token rings, FDDI nets, fiber rings, and satellite networks. Private networks within the context of a global enterprise network are more likely to use a variety of wire and wireless transmission systems to achieve mobility, flexibility, remote interconnection, and lowest cost of service.

The attempt over the last 20 years to remove corporate networks as far as possible from the regulatory oversight imposed on the private switched telecommunications networks makes it difficult to regulate and control these private networks today. The remarkable growth of these networks from being almost nonexistent to an ever-growing percentage of all telecommunications networks is clearly redefining their nature and their overall importance. Further, some private networks such as ANPs, CAPs, and teleports function in almost every way like a public carrier except they are at least in name acting as a "nonpublic" system. In the United States this rapid growth in private networks, out-sourced private networks operated by third parties such as EDS or Anderson Consulting, or virtual private networks provided by carriers have all been seen as the logical result of a competitive marketplace on one hand and the desirability of being nonregulated on the other.[18]

In Europe, the European Commission has backed open competition in the field of telecommunications. In particular, it has through the efforts of its Directorate IV on competition as well as its Directorates on telecommunications and space activities (Directorates XII and XIII, respectively) consistently promoted open access to public telecommunications networks by value-added carriers, authorization of private satellite networks, and more opportunity for competition to the public switched telecommunications networks. Despite this pressure from the offices of the European Commission in Brussels, acceptance of open competition, private networks, licensing of new wireless systems, or approval of unlicensed frequencies for wireless networking has been uneven throughout Europe and probably a decade behind the United States in terms of official national governmental level support.[19]

Japan has also moved to allow competition among new telecommunications carriers for long-distance and international services, but has been slow, presumably for national strategic and security reasons. The Japanese Ministry of Posts and Telecommunications, for instance, has not supported private networks that are other than value-added carriers. It has supported a form of "tri-opoly" by creating three competitive units to install networks for national long-distance and overseas telecommunications. All in all, the global picture is a mosaic of diverse actions. Some developing countries have allowed the installation of new wireless networks in specific geographic locations and for specific services. Some have allowed interconnection to the PSTN and others have not. In all some 50 countries have moved to use competition as a major market regulatory tool for their respective overall telecommunications frameworks. Most of these countries are members of the OECD and the nature and extent of deregulation and market competition varies widely around the world.

Altogether, as of the end of 1994, about 50 countries around the world allow the creation and operation of private networks for mobile and fixed telecommunications services, but each country has its own rules and regulations. Some allow only wireless systems, some allow only wire- or cable-based systems and some allow both. Some countries allow relatively easy entry but others require direct governmental or local investment. Some even require a majority share of the investment. Global corporations actually now have a wide range of choice. They have the opportunity of establishing some form of wireless or wire-based private networks or competitive access system in some 50 countries. This means that many corporations will, in fact, choose where to locate plants or regional offices on the basis of the ability to create private communications networks within the host countries. Alternatively, this can mean that countries with a single monopolized telecommunications carrier and prohibitions against any form of private networks even in the form of a value-added carrier may be carefully avoided by new investors.[20]

There are, of course, other problems that go beyond the control of private telecommunications networks. One of the key regulatory issues is how to address the problem of official governmental use of wireless frequencies. There are rather different types of uses that can conflict with one another and therefore require a systematic approach to achieve an integrated utilization plan. A privatized and competitive environment obviously requires a well-coordinated plan and a carefully managed system of frequency allocations. In cases where several governmental agencies or the defense ministry controls frequency allocations, there can be problems of conflicting frequency allocations.

There are military- and defense-related requirements that involve dozens of different applications and frequency bands and often involve the whole country or even the entire world, such as military telecommunications satellite systems or radar operations. At the opposite extreme are police, fire, public safety, and local governmental operations that involve a limited range of frequencies and specific geographic areas. Finally, there are frequency allocations for national governmental operations that can range from national forestry management, to map-making, to emergency warning networks. The fact that governmental- and military-use frequency allocations and assignments that are made by different agencies and at different levels (i.e., national, state, or provincial and local levels or military versus civil) often lead to inefficient and even rather haphazard patterns of use.

If one adds commercial carriers of wireless services on top of this already difficult environment, the problems can become complex indeed. Multiple private, commercial, and governmental users such as is the case in the United States can create multiple sources of interference on one hand or can lead to highly inefficient use of spectra on the other. Often there is no appeal method whereby antiquated, underutilized, or ineffective frequency uses by governmental agencies, the military, or both can be effectively reviewed. Recent attempts to reallocate and auction off underutilized bands in the United States suggest, however, that a new era of reexamination of past practices may be underway. In time this may spread to other countries as well.[21]

7.2.3 Specific Regulatory Authorizations: Construction Permits and Licenses for the Provision of Wireless Service

To operate a wireless service in the United States also requires the use of radio wave spectra. This typically means fulfilling a number of regulatory steps. These include obtaining a license that specifies the frequency band, power-level limits, safety provisions, and other constraints. It also often involves a construction permit to build the proposed system. Further, if the system is to operate as a common carrier rather than as a private network then it may require the filing and approval of tariffs as well. In the case of most mobile services the specified frequency bands for the precise service is a fixed and well-defined resource that may not change for many years.

While new allocations do occur, the 200 MHz of frequency in the 2-GHz range newly assigned to PCS service is the exception rather than the rule. Such new allocations are actually rare events that typically will occur no more often than once a decade or even less frequently. Further, the FCC-controlled process has tended to consolidate the assignment of new frequencies into a single macroevent to cover all allocations for the country in a single, consolidated, and competitive process. In the case of the PCS allocations, however, the process for the United States was different from past procedures in that the frequencies were auctioned off to the highest bidder.

In the case of the cellular licenses assigned in the mid-1980s, the FCC provided one license to the local telephone company and then had a lottery for the other competitive cellular license assigned in every major urban market. In the case of satellite systems, the FCC held comparative hearings. Most recently, the FCC was authorized by the 1993 Comprehensive Budget Act simply to auction off the frequency bands for the PCS service. The idea in this case was to raise revenues for the federal government and to pay for the cost of transition to new frequency bands for the current users of the bands. Despite the three different approaches that have been recently applied to frequency allocations and the awarding of licenses, (i.e., lottery, comparative hearings, or auction), the FCC has tended in all cases to at least be consistent. In each case the FCC has followed a similar basic process. This has been to identify the spectra resources available for a new or expanded service, to notify the commercial organizations who might be interested in providing this new offering, and then to create a single process and clear-cut set of rules for deciding who would be chosen and who would not. What is rather fundamentally different about PCS licenses is that far more licenses are being given out and the competitive framework that is being created will likely be more than the market can reasonably sustain. Some large telecommunications carriers may in effect wait for market failures and then buy into the PCS at "discount" rates.[22]

The only exceptions to this systematic process of choosing the winners and losers has been the limited use of negotiated rule-making wherein the affected and interested parties seek a mutually agreed-upon solution.[23] There is also the relatively new process wherein the FCC grants a "pioneer preference" for new technical innovators. The three pioneer licenses for PCS, for instance, were granted to Cox Cable, the Washington Post Corporation, and Omnipoint. The pioneer preference was first proposed by former NTIA head Henry Geller and have proved to be highly contentious. The intent of pioneer preference status was to stimulate innovation, especially among new and small entrants. It was also intended to speed up frequency allocations, accelerate licensing for innovative applicants, and resolve intellectual property rights issues. In practice, the awards have seemed to have been made in an inconsistent manner and have led to court challenges of FCC actions and created confusion among applicants as to what to expect from their applications. Some believe that the court challenges and litigious results from this process will lead to it being discontinued.[24]

7.2.4 The Increasingly Popular Unlicensed Wireless Industry— Part 15 or IMS Frequencies

The allocation of frequencies for personal communications services and digital cellular services in the United States has progressed rapidly in the last few years from a very modest activity with very limited spectra available for these services to a major mass-market service that is producing billions of dollars in revenue for mobile service providers willing to vie for exclusive licenses. The precursor market that foreshadowed this remarkable new growth of mobile services was a phenomena of the 1960s and 1970s known as Citizens' Band Radio. This almost automatically self-licensing service, which provides a rather limited-quality push-to-talk communications service for cars, trucks, and recreation vehicles demonstrated that a large number of users could indeed share a limited number of nonexclusively assigned channels.

Today, many suppliers of radio transceivers have decided to opt for selling equipment designed to operate in nonexclusive Industrial and Medical Service (IMS) bands allocated for short-range applications. This allows manufacturers to build and sell wireless data modems, wireless PABXs, or wireless LANs in an open use, nonproprietary, nonexclusive use frequency band. Carriers that have broad band fiber networks available to them for longhaul transmission can actually operate wireless mobile networks using the nonlicensed frequencies at the ends of the network. This is particularly successful in a geographically delimited corporate campus site or research complex with a single telecommunications management entity. It can also be used for noncontending data transmission over a much larger area if real-time voice communications is not required. Some portable computer manufacturers have used these Part 15 bands to connect their computer to office LANs or to laser printers. This sometimes overlooked market is now emerging as a multibillion market.

From a policy viewpoint, access and use of IMS Part 15 bands is rather straightforward. Regulations are set as to frequency bands, channelization, and permissible power levels. The suppliers, through conformance testing, largely police their own activities. The FCC retains the right to fine or penalize transgressors of the rules. As the scope and volume of use increases in coming years, this area could, however, become much more unmanageable to monitor and control.[25]

7.2.5 Social and Public Policy Issues from Wireless Communications Services

There are actually a number of cases where cellular or portable telecommunications services give rise to special problems involving social issues. These include problems of emergency 911 services through cellular phones, the issue of mobile services for the handicapped, and special problems areas where mobile services and criminal activities can become linked.

There are a number of perceptions about mobile cellular services that are either wrong or somewhat distorted. One of these perceptions is that the world of mobile cellular telephone service is seamlessly connected to the terrestrial-wire PSTN network and that essentially all features you can obtain on the terrestrial-wire PSTN network are likewise available as a mobile service even though you may have to pay additionally for it. The truth is that the limited-channel and signaling capabilities of cellular systems do prevent the provision of certain mobile services today. Many of these can be considered discretionary in nature and involve only slight deficiencies in service quality or throughput. With PCS service a number of the limitations will be overcome.

In one area, however, mobile telephone service deficiencies represent a key social policy issue. This is with regard to the reliability and functionality of "911" emergency services.

Many of the cellular systems implemented in the United States were hurriedly installed to begin producing revenue as soon as possible. The results, at least in a number of instances, were cellular systems that inadequately handled 911 calls. First, there are a number of cellular systems where one must dial a number other than 911 for emergency service. Second, there are a number of systems where there can be misidentification of the originating switch for tele-location purposes. There have been a number of cases of "mistargeted" cellular emergency 911 calls. In one case of a gang attack the originating call came from the southern section of Denver but was identified as Ft. Collins, Colorado, some 60 miles away. In Houston a pro football player who sought help after seeking to commit suicide could not be tele-located at all. This is not to suggest that tele-location services are not generally available or do not work on cellular systems. It does suggest that there are many instances where this is true and that enforcement provision in the area of cellular 911 services need to be uniform and strictly enforced.[26]

In well-equipped mobile systems the signaling information indicates the exact cellular antenna that is handling every call. It should indicate not only the cellular antenna that is handling the call (i.e., Mobile Telecommunications Switching Office Number 12, Cell 38 North antenna), but also when each hand-off occurs. This type of standardized approach could help to pinpoint the appropriate location of the mobile subscriber to within one-quarter of a mile of their emergency call.

In the new digital cellular mobile telecommunications systems and the new Personal Communications Services, the enhanced 911 services can be greatly improved to provide faster and higher-quality emergency responses to 911 calls. Databases supporting 911 can be augmented to store information about medical problems and key health conditions, allergies to specific drugs, and so on. It is also possible to upgrade ranging information to exactly pinpoint mobile locations. This can be accomplished with LORAN-C or even satellite-based positioning using GPS receivers for subscribers willing to pay for such service. In this case one could subscribe to a satellite-based GPS for navigational purposes but would obtain emergency tele-location services as an additional benefit. In all of these cases, the technology is largely available today. The issue is thus one of mandatory standards for enhanced 911 on mobile service systems. The GPS tele-location service would, of course, be optional since the cost of this service is currently too high for universal implementation, even on mobile systems.[27]

There is another policy issue to be considered here, however, and that is the issue of privacy and confidentiality. The ability to pinpoint a precise location and to have access to someone's medical records is good in the case of a serious automobile accident or, say, when someone is being attacked. However, most individuals would not want this information disclosed for other purposes. The problem is a question of how one ensures that an emergency information system is not accessed surreptitiously for other than the intended purposes. A combination of measures are likely needed. These would include carefully administered and updated security access codes, severe criminal penalties for improper use, and third-party audits of the uses of the enhanced 911 system.[28]

Another key area where mobile communications services give rise to policy concerns is that of handicapped services. In the United States about 8 percent of the population is in one sense or another physically handicapped. Individuals who are handicapped with regard to speaking, hearing, seeing, or walking, including those who are subject to attacks that may temporarily limit their abilities, have special needs. Significant progress has been made in recent years to create special communications equipment and tele-alarm systems for the handicapped, both within the home as well as in the workplace. The provision of handicapped services in the form of mobile communications systems have been much more limited. Mobile tele-alarm systems for personal cars, buses, trains, airplanes, and airports are needed to allow handicapped individuals to function normally in society and to hold jobs.[29]

There are many unresolved policy issues in this respect. What telecommunications services should be mandated for mobile services to the handicapped and who should pay for them? Should certain basic services be subsidized by all subscribers, for all mobile subscribers, or just by the handicapped who might use them? Are there advanced mobile services for the handicapped such as visual displays for the deaf, or special keyboard units for those without speech that would be only partially subsidized or perhaps not subsidized at all? Most of these issues of service to the handicapped have not been resolved in terms of legislation but will likely be addressed in new federal legislation in coming years. Part of this discussion will focus on the issue of at what time does mobile service stop being an optional and "high-value" consumer offering available only to a few and when does it start being a basic or essential service?[30]

A final key issue of importance with regard to mobile communications service is that of what might be called "tele-crime." The truth of the matter is that mobile and wireless telecommunications of various types have helped to facilitate certain forms of criminal behavior. Although almost any telecommunications system can be eavesdropped upon, the easiest way to monitor calls is tune in on an analog wireless call. All that is needed is a simple scanner than is tuned to the precise frequency in use and to listen in on the conversation typically from a 100 to 200 foot radius. This ability to listen in on portable or cellular phones in use in a house or backyard or even in a car can facilitate robberies, credit card fraud, kidnappings, industrial sabotage or espionage, or even blackmail and extortion. The move to digital telephony and the advent of the clipper-chip-like technology can serve to diminish these problems, but clearly this is a major problem of 1990s style crime.[31]

Cellular telephones also give rise to additional concerns with regard to criminal behavior. One serious problem is that of stealing an automobile with a cellular phone and then using it to make expensive overseas calls. This can often be related to criminal behavior such as in the case of making drug deals. It is then easy to abandon the car and start the process over again. This allows thousands of dollars of calls to be made fraudulently and it also permits originating calls from an anonymous and nontraceable source. In the case of drug deals, laundering of money, and other such transactions, this type of cellular telephone fraud has become commonplace. The need for some form of countermeasures against the misuse of cellular phones has received increasing attention. Some have suggested that similar systems as those planned for the American Mobile Satellite Corporation national mobile telecommunications service should be extended to all cellular systems. In the AMSC system a code must be given before an equipped automobile will start and if the wrong code is given repeatedly an alarm is given along with tele-location information automatically going to the nearest police facility. Further, the car and the telephone set are deactivated. This could be a serious problem for inebriated owners who forget their proper code.[32]

This is not an exhaustive list of public policy issues associated with mobile telecommunications services, but problems with mobile-based enhanced 911, mobile services for the handicapped, and with criminal behavior are indicative of the problems that exist. In many cases new technology or applications can help, but there is also the question of who pays and how much and how soon. In other cases of illegal activities, clearer or stiffer penalties against misuse are needed. The enactment of a new U.S. federal law to impose stiff penalties on those who use scanners to eavesdrop on and tape conversations off cellular or satellite conversations as of October 25, 1994 is certainly a step in the right direction. As mobile communications spread and become more and more pervasive throughout society, these issues will need and likely receive increased attention. Presumably this will eventually lead to corrective action being taken at the level of new legislation, societal education, or police enforcement. In the United States, where a combination of state and federal laws plus state and federal regulatory commissions decide what is right and wrong and what services can be provided at what cost and under what regulations, the telecommunications environment can be not only confusing, but even contradictory.

7.2.6 Important New U.S. Initiatives to Deregulate Mobile Communications

Today, the regulatory environment for telecommunications in general and wireless services, in particular, are in a state of rapid flux. There are several bills in Congress that are pending action. Despite the lack of action on major telecommunications legislation in 1994, there still seems to be a willingness to introduce further competition, especially at the logical loop and across traditional lines of business. Some of the major changes being sought through legislation would change the strategic balance and relationship between the U.S. interexchange carriers (e.g., AT&T, MCI, Sprint, Wiltel, etc.) and the regional Bell operating companies (e.g., Ameritech, Bell Atlantic, Bell South, NYNEX, Pacific Telesis, Southwestern Bell, and US WEST). In particular, these new laws would largely dismantle the Modified Final Judgment of Judge Harold Greene in the following ways. First, the interexchange carriers, as well as cable television and alternative network providers, would be able to compete for local telephone services. Second, the local exchange carriers would be allowed to compete for long distance and international traffic. Third, the Bell operating companies would be able to compete for cable television services in their local areas and to provide both programming and transmission capabilities. Major legislative reforms in telecommunications have been predicted for many years. The truth is that the last comprehensive legislation was enacted some 60 years ago, well before the advent of satellites, cable television, cellular, or PCS. Only time will tell if Congress can indeed update telecommunications legislation to bring it into the 1990s and form a comprehensive regulatory framework suited to the twenty-first century.

In some of the measures the wireless industry would also be further opened to competition. If these rather sweeping changes in the U.S. regulatory environment were to take place it would imply a major upset in wireless communications. One of the main factors that pull interexchange carriers into mobile communications is direct access to end-user customers so as to "bypass" local exchange carriers and their high access charges. As noted earlier, the AT&T multibillion take over of McCaw Communications is clearly driven by the desire to reestablish direct contact to the individual subscriber.[33]

If, on the other hand, interexchange carriers were suddenly to be able to compete in the local exchange market, the strategic importance of mobile services is likely to be diminished. In short, if interexchange carriers were to be fully empowered to acquire Alternate Network Providers like Metropolitan Fiber, or purchase and operate cable television systems in any market, or if they could install their own fiber networks to serve business or selected residential areas, they would be freed to pick and choose between "wire" and "wireless" technology in new and innovative ways not open to them today.

As of early 1995, the interexchange carriers have only limited access to the end-user, (except for the four out of 50 states that allow interconnection directly at the local loop level). They can use ANP's to link businesses to their point of presence or they can connect through wireless networks. The multibillion dollar price that AT&T paid for the acquisition of McCaw Communications is just one indication of how important "bypassing" the access fees of the local networks actually is. Another more quantitative and exact way of stating this importance is that under current tariff structures the interexchange carriers pay up to 45 percent of their revenues to access charges that go to local exchange carriers.[34]

The regulatory environment in the United States with regard to competition in the local loop will change dramatically one way or another in the next few years. Today there are at least three legislative proposals in the U.S. Congress to allow more competition in the field of telecommunications. These largely pertain to equipment manufacture by telecommunications carriers and to competition in the local loop among and between interexchange carriers, local exchange carriers, and cable television companies. Although other industries may not be explicitly addressed one can also anticipate that power utility companies, electronic publishers, broadcasters, the entertainment industry, and publishing houses could also join the competitive fray in terms of both new infrastructure and content for the so-called information highway.

If one were to interpret the implications of these changes for the wireless telecommunications business, it is possible to project both positive and negative trends. On one hand, if wireless services become less important as a bypass mechanism between the consumer and interexchange carriers, this might be seen as a deterrent to future rapid growth. Alternatively, if more competition is introduced at the local loop, this could stimulate efforts to offer seamless and lower-cost hybrid services in terms of integrated wireless and wire services to the consumer.

One might speculate that both business and the residential consumer would opt for the "best" and most cost effective "package" of telephone, fax, television entertainment, and mobile communications services. This scenario would suggest that everyone is involved— the cable television people, the local exchange carriers, the interexchange carriers, the electronic appliance and computer vendors, the software giants, the power companies, the DBS providers, and even electronic publishing industries. All of these industrial groups are seeking to offer "the complete information highway package" that would include among many other services the popular and highly profitable value-added wireless services. The actual outcome in this respect, however, is still some years away. It is likewise too early to indicate what will be the optimum hybrid mix of wire and wireless technologies that will make these offerings possible. The fact that it will be mix of both technologies, however, seems quite clear.[35]

Finally, it should be noted that apart from any further regulatory initiatives, the move to create a new fabric of increased competition through the issuance of a huge number of PCS licenses, the authorization of a sizable number of new satellite systems, and the encouragement of unlicensed wireless services is creating fundamental regulatory change. The use of competition as a regulatory mechanism is reaching new extremes. Some feel that the creation of multiple standards and the lack of sufficiently clear interconnection guidelines could ultimately spell future trouble to smooth system operations in the twenty-first century. Only time will tell how much competition and how many competitors are best for the consumer.

7.2.7 World Regulatory Environment—A Comparative Look at the United States and the Rest of the World

The world at times seems to view the United States, at least in the field of telecommunications, with a mixture of awe, disbelief, amusement, experimental curiosity, horror, and a large dose of skepticism. The world of communications certainly considered AT&T prior to the 1984 divestiture to be one of the best telecommunications systems and best-managed corporations on the globe. The sentiments often expressed within the general citizenry of the United States at the time of the breakup of AT&T was first being implemented were skepticism and incredulity. Why do it? The rhetorical question that echoed around the world was: "If it isn't broken, why fix it?" After more than a decade of experience with the competitive model within the United States, there is much more known about the benefits and liabilities of an open and competitive market. The values of competition in terms of reducing prices, stimulating innovation, and reducing subsidy levels are much better understood, but the problems of fraudulent practices, developing needed interconnection standards, and ensuring of network-wide compatibility remain key issues of concern.

The significant power of governmental policy and regulatory control of telecommunications markets should never be underestimated. Technology, applications, service offerings, standards, and tariffs are all important but ultimately the policy makers have the power to redefine almost everything. They can legislate, arbitrate, adjudicate, regulate, restructure, and even confiscate. They are often the gate keepers to key resources like frequencies, right of ways, market access, and service offerings. Despite the complexities, the lengthy and cumbersome quasi-legal procedures, and sometimes inconsistency of telecommunications policies, the U.S. system is still among the world's most open, most consumer-oriented and fairest regulatory regimes in the world. U.S. regulatory leadership, which has defined a framework of open competitive networks and rapid innovations in digital technologies, has been followed by a growing number of countries around the world.

The telecommunications corporations who are often the most adept at coping with a changing and challenging regulatory environment are also those that are among the best organized and managed. Good management is, in fact, often a combination of good planning, good structure and personnel, and a well-conceived basic philosophy. This is often a commitment to consistent goal-driven activities devoted to a high-level mission rather than short-term opportunism and narrow regulatory maneuvering.[36]

In short, success in the world of regulatory policy is often tied to effective and mission-driven management and *vice versa*. Good regulatory strategies and good corporate management are frequently joined together. At their best, they share clear-cut and longer-term goals, good and effective communications, competition in the marketplace of ideas, and commitment to ideals such as quality, fairness, and letting new ideas surface and prosper.

The world of telecommunications is likely to experience on-going regulatory change for some time to come. Most of the countries within the OECD have now created new regulatory frameworks in the field of telecommunications to allow more competitive markets, restrict or eliminate most *de jure* monopolies for telecommunications, allow international access to previously restricted national telecommunications markets and allow the operation of value-added networks as well as virtual private networks. This is a broad and sweeping change that is also being considered in a growing number of industrializing and developing countries. Today, however, a second major step is beginning.[37]

This new step, like the earlier steps that began in the 1980s, also started in the United States. This new telecommunications regime is being created as much by new technology as it is by any regulatory, judicial or legislative change—even though it combines all of these elements. This is the move toward the so-called information highway and the convergence of the communications, computer, content, and consumer electronics industries. It is also being led by software-defined networks that are allowing corporate communications systems, especially so-called enterprise networks, to provide all forms of services—voice, data, fax, e-mail, video-conferencing, imaging, scientific visualization, interactive CAD/CAM, multimedia, and even virtual reality. The old divisions that once separated voice and data, telecommunications and LANs, long-distance and local service are all evaporating.[38]

The basic framework that has been identified by the Clinton Administration for the implementation of the National Information Infrastructure (NII) has been reasonably well defined within the context of nine explicitly defined principles. These have been set forth as follows[39]:

(a) promote private sector investment;
(b) extend the universal service concept to ensure that information resources are available to all at affordable prices;
(c) promote technological innovation and new applications;
(d) promote seamless, interactive, user-driven operation;

    (e) ensure information security and network reliability;

    (f) improve management of the radio frequency spectra;

    (g) protect intellectual property rights;

    (h) coordinate with other levels of government and with other nations;

    (i) provide access to government information and improve government procurement.

Although these principles are quite broad and are in the process of being further refined, there is a clear and comprehensive framework under which NII implementation can proceed. The references to frequency spectra efficiency; to seamless, interactive, and user-driven applications; to universal access; and to technological innovation would all seem to support the idea that both wire and wireless technologies will be needed to achieve these goals. In short, these goals do not seem achievable by only using terrestrial fiber optic technology. Instead a hybrid network of fixed and mobile services and technologies seem most likely to achieve the desired results.

The concept of the NII is leading to a great deal of change and even chaos. Old borders between industries are fast disappearing. The distinctions between broadcasting and telecommunications common carriers are largely gone. Even the structure of the FCC, as based on the industry structure of the past, no longer makes a great deal of sense. Certainly, the legislation that has controlled the field of communications since 1934 is largely passé if not obsolete. Within five years this same problem, which can be summarized as "digital convergence," will spread to the countries of the OECD and then the rest of the world.

The ripples of regulatory change will reverberate around the world. Within all this change, wireless communications will be only a part of the broader transitions occurring in the field. Wireless will nevertheless be an important part of the coming innovations. This is particularly true because radio and mobile wireless technologies can be deployed and implemented much more rapidly than wire- and fiber-based systems. Therefore the impact can be seen much sooner, say in millimeter wave cellular terrestrial systems (i.e., LMDS) or in new satellite systems that might be deployed for entertainment, mobile, or other new telecommunications systems.[40]

Some critics have claimed that the U.S. regulatory process, which is now so very heavily based upon competitive practices, needs to reform its regulatory procedures within the FCC to place more emphasis on ensuring a "fair and level playing field" for competitors and spend less time on monitoring of tariffs and industry processes. They note that the new competitive regulator mechanisms created in the United Kingdom and Australia (i.e., OFTEL and AUSTEL) are able to monitor citizens' and industry complaints and impose stiff fines to enforce fair competition and can do so with only a small fraction of the personnel and costs of the FCC. In the case of the United Kingdom, solid enforcement and heavy fines have served the cause of fair competition and this with an OFTEL staff of under 100 people. Even if one acknowledges that the United States is five times larger than the United Kingdom, OFTEL is in relative terms at least three times smaller than the FCC. Each country has its own telecommunications concerns and priorities, but the merits of this more streamlined approach may well be considered in future years.

7.3 MODERN MANAGEMENT CONCEPTS AND THE WIRELESS TELECOMMUNICATIONS INDUSTRY

In the world of modern management there are several key new theories of management that tend to put market and customer responsiveness first in overall priority. Few providers in the highly competitive wireless telecommunications field will succeed in the twenty-first century. Even fewer will succeed without a commitment to consumer-oriented management philosophies. Two of these seem particularly worthy of note. One of these is known as Time Based Competition (sometimes called Time Based Management) and the other is known as Total Quality Management. To understand the critical role of customer-responsive management, it is useful to note and compare the critical aspects of both of these key theories of modern management as given below.

The basic tenets of Time Based Management are given in eleven key points[41], as follows.

Key Elements of Time Based Management
 (1) Review and revise corporate processes.
 (2) Analyze and amend *all* functions, not just some.
 (3) Upgrade standards and achieve better quality.
 (4) Create teams and teamwork.
 (5) Improve communications.
 (6) Track time and speed as much as money.
 (7) Create a corporate culture that respects time and responsiveness.
 (8) Enhance scheduling and information technology systems.
 (9) Streamline approval processes.
(10) Restructure and reorganize to accomplish TBM objectives.
(11) Emphasize education and training.

The key element of Total Quality Management as developed by Dr. W. Edwards Deming, the so-called "father" of modern industrial Japan, are presented in a total of 14 points, which if analyzed closely actually have much in common with the key elements of Time Based Management.[42]

Key Elements of Total Quality Management
 (1) Create constancy of purpose.
 (2) Adopt the new philosophy.
 (3) Cease dependence on inspection to achieve quality.
 (4) Cease doing business on the basis of prices alone.
 (5) Improve constantly and forever the system of production and service.
 (6) Institute training on the job.
 (7) Institute leadership.
 (8) Drive out fear and fault-finding from the workplace.
 (9) Break down barriers between departments.
(10) Eliminate slogans and targets as management philosophy.
(11) Eliminate numerical quotas.
(12) Allow pride in work.

(13) Institute a program of self improvement.

(14) Involve everyone in the organization in achieving the transformation.

Although the words and concepts as presented seem different, there are in fact many aspects that are very similar. These two management philosophies share the following common points:

- Developing a great respect for "quality" rather than inspection.
- Instilling teamwork, good communications, and shared or common purpose.
- Eliminating unnecessary overhead, streamlining levels of administrative bureaucracy, and eliminating structural and administrative roadblocks where they occur.
- Respecting people and emphasizing training and education.
- Doing things "right" and "better," rather than by rote or by arbitrary quotas.
- Realizing that the only good management is an improving management.

This commonalty is really incidental to an even more basic linkage and that is that both TBM and TQM are in a large part built upon creating a corporate management that is based upon what the customers' of that corporation would most likely want in terms of being able to obtain the best possible products or services at the lowest possible cost. In the highly competitive and fast-growing world of wireless telecommunications there is a good deal of evidence that a successful corporation must adopt these strategies or will very likely fail. Some of the most successful corporations in this field such as Motorola, Northern Telecom Inc., and Nippon Electric Corporation, to name only a few, are world leaders in the implementation of these management concepts.[43]

The "point" of both management systems is thus ultimately to respond to customer and market demand in effective, interactive, and highly responsive ways. There is, of course, the opposite corollary to this approach to management. This is to say that many corporations are designed, structured, and managed to benefit not the customer, but rather the top and sometimes the middle management. Customers, for instance, would most likely want highly trained employees, streamlined management, a strong commitment to quality and value, good communications, teamwork, and effective and innovative response to consumer demand. They would probably tend not to want high or excessive employee and executive salaries, complicated approval cycles, highly stratified layers of management and excessive staffing, slow product development, slow rates of innovation, large and expensive offices, long vacations and excessive "perks," or indifference to customer input and thoughts.

In Chapter 1, the concept of the five drivers in telecommunications and the marketplace was presented. One might argue that the representation of this model is upside down. At least in priority, the market and consumer demand should be at the top of the chart and the five key components of change and innovation should be shown as subservient to the market forces. This would suggest that the majority of the key changes in telecommunications should start with the market. The wireless services that are especially market driven would seem to be especially so.[44]

The first key to understanding and responding to a market is a management approach that gives this objective top priority. Clearly there are unique aspects of the wireless market that should be considered on a segment-by-segment basis, but the basic fact is that a badly managed organization with inadequate means of assessing market demand and consumer needs will not succeed regardless of the product or service being offered. Nine times out of ten a well-managed corporation with a good longer-term strategic plan, and a well-conceived customer-oriented market plan can effectively operate and even succeed in even a shifting regulatory environment. This does not negate the need for good legal advice, strong participation in standards and governmental advisory groups, and constant attention to key policy and regulatory issues as well. Ultimately, balance is needed in both management and regulatory policy.

ENDNOTES

(1) Lucas, J., "PCS vs. Cellular," *Telestrategies*, August 1993, pp. 1-12; also see Holland, B., "Wireless Telecommunications, PCS, and In-Building Services" (Broomfield, CO: Spectrallink, 1994).

(2) The regulatory and standards issues involving frequency allocations are addressed separately in Chapters 2, 6, and 8.

(3) See Chapter 5 for detailed discussion of these matters.

(4) "Evolving Wireless Charging Policies," Wireless Technology Conference (Washington, DC: Phillips Publishing, September 1993), pp. 21-23.

(5) "Evolving Wireless Charging Policies," Wireless Technology Conference (Washington, DC: Phillips Publishing, September 1993), pp. 26-28.

(6) Pelton, J.N., *Future Talk* (Boulder, CO: Cross Communications, 1992).

(7) AT&T's McCaw Deal Offers Bypass of RBOC's Access Charges," *Wall Street Journal*, April 12, 1993.

(8) Alleman, J., "The Advantages of Universal Measured Service vs. Basic Flat Rate" (Sydney, Australia: International Telecommunications Society, June 1994); also see "Universal Measured Service," *GTE Technical Staff Paper*, Stamford, CT, 1991.

(9) "PCS Bidding Rules Issued: Messaging Services To Be Licensed This Summer," *Inside Wireless*, April 27, 1994, Vol. 2, Issue 5, pp. 1-2.

(10) Lucas, J., "PCS vs. Cellular," *Telestrategies*, August 1993, pp. 5-12.

(11) Schnee, V., "NEXTEL's ESMR Arrives—A New Era in Cellular Integrated Services," *Wireless*, January/February, 1994, Vol. 3, No. 1.

(12) Gorman, B., "Enterprise Networks and Corporate WANS: The DEC Experience," ICA Annual Conference, Dallas, TX, May 1992, pp. 113-115.

(13) "International Computer Telephony Association Formed," *Telephony Magazine*, June 1993, pp. 12-14.

(14) Gorman, B., "Enterprise Networks and Corporate WANS: The DEC Experience," ICA Annual Conference, Dallas, TX, May 1992, pp. 115-117.

(15) "Current Efficiency of Use of Radio Spectrum," U.S. Government Accounting Office, Staff Report, April 1992.

(16) Gorman, B., "Enterprise Networks and Corporate WANS: The DEC Experience," ICA Annual Conference, Dallas, TX, May 1992, pp. 117-121; also see Special Issue on Enterprise Networks: *Telematics and Informatics*, April 1993, pp. 12-16.

(17) "Competitive Access and Bypass in International and National Telecommunications Networks," INTUG, February 1992.

(18) Special Issue on Enterprise Networks: *Telematics and Informatics*, April 1993, p. 14.

(19) "Virtual Private Networks and Value Added Networks," European Commission, Directorate XII, Staff Paper, March 1992, pp. 24-28.

(20) "Type II Carriers and Value Added Networks," Ministry of Post and Telecommunications, Tokyo, Japan, December 1992.

(21) "The Future of Radio Spectrum Management," *Telecommunications Policy* (London, U.K.: Butterworth-Heinnemann, 1992).

(22) "New Applications in Wireless Services," FCC Staff Report, March 1993.

(23) "Negotiated Rule Making for Land Mobile Satellite Services," FCC, Washington, DC, January 1993.

(24) FCC Radio Regulations, "Application Procedures for Pioneer Preference," FCC, Washington, DC, 1993.

(25) Holland, B., "The Unlicensed Wireless Telecommunications Market" (Broomfield, CO: Spectrallink, 1994).

(26) "Enhanced 911 Services: Opportunities and Problems," SCC Report, Boulder, CO, 1994, pp. 2-6.

(27) "Enhanced 911 Services: Opportunities and Problems," SCC Report, Boulder, CO, 1994, pp. 7-8.

(28) "Enhanced 911 Services: Opportunities and Problems," SCC Report, Boulder, CO, 1994, p. 10.

(29) "The American Mobile Satellite Corporation (AMSC) and New Mobile Vehicular Services," The Satellite Users Conference, Washington, DC, September 1994.

(30) "The American Mobile Satellite Corporation (AMSC) and New Mobile Vehicular Services," The Satellite Users Conference, Washington, DC, September 1994.

(31) "Mobilecom Services," *GPS Report*, April 1994, pp. 3-5.

(32) "Mobilecom Services," *GPS Report*, April 1994, pp. 4-6.

(33) "AT&T and Other IXC's Wireless Strategy," *Telecommunications*, February 1994.

(34) Lucas, J., "Local Loop Competition," *Telestrategies*, May 1993, pp. 1-3.

(35) U.S. Congressional Bill HR. 3636, 1994.

(36) Bender, G., "White Paper on Telecommunications," September 1994.

(37) "Virtual Private Networks and Value Added Networks," European Commission, Directorate XII, Staff Paper, March 1992, pp. 24-28.

(38) Electronic Industries Association and Telecommunications Industries Association, "White Paper on the National Information Infrastructure," Washington, DC, June 1994, pp. 1-4.

(39) Electronic Industries Association and Telecommunications Industries Association, "White Paper on the National Information Infrastructure," Washington, DC, June 1994, pp. 5-6.

(40) Pelton, J.N., "Convergence of VPN, Enterprise Networks, PSTN and Cable TV," Lecture, Interdisciplinary Telecommunications Program, University of Colorado, April 1994.

(41) Lawton, B., "Time Based Management," Seminar Lecture, Interdisciplinary Telecommunications Program, University of Colorado, September 1994, pp. 1-10.

(42) Deming, W.E., "Principles of Total Quality Management," *Management Principles*, 1985, pp. 7-11.

(43) Lawton, B., "Time Based Management," Seminar Lecture, Interdisciplinary Telecommunications Program, University of Colorado, September 1994, pp. 1-6.

(44) "The Globalization of Universal Telecommunications Services," in *Universal Telephone Service: Ready for the 21st Century* (Wye, MD: Institute for Information Studies); *Annual Review*, Aspen Institute, November 1991, pp. 141-151.

CHAPTER EIGHT

STANDARDS FOR WIRELESS AND SATELLITE SERVICES

8.1 INTRODUCTION TO THE STANDARDS-MAKING PROCESS

The standards for wireless and satellite communications have grown almost exponentially in recent years. This trend toward multiple standards seems likely to continue for a number of years into the future before reversing. This rather chaotic standards environment has at least six identifiable causes. The key factors that serve to decentralize and proliferate standards include those shown in Figure 8.1.

Factor A:	The difficulty and expense of obtaining access to scarce frequency bands. This shortage can lead to using "new" services and or "new standards" as a means of obtaining frequency allocations.
Factor B:	Differences in viewpoints concerning the effectiveness and overall utility of different modulation, encoding, or multiplexing techniques.
Factor C:	Differences concerning other key technical factors such as optimum cell coverage areas, power levels, methods and techniques for node interconnection, degree of needed interactivity, determining the optimum frequency band, and the necessity for integrated positioning, paging, data, and voice services.
Factor D:	Differences concerning service offerings, market size, and the viability and profitability of different mobile services.
Factor E:	Industrial interests that dominate certain existing standards in different geographic regions of the world.
Factor F:	The overall trend in telecommunications standards toward regionalization and the diminished importance of the global standards process.

FIGURE 8.1 Reasons for the Proliferation of Mobile Telecommunications Standards

Today, in the United States alone, there are some 94,000 standards being actively maintained by the American National Standards Institute (ANSI), the National Institute of Standards and Technology (NIST), the U.S. Department of Defense, and so on. Of these U.S. standards, over half have been developed under the auspices of the federal government and the rest by the private sector. For the field of wireless telecommunications the entity with probably the most important influence and overall power is the American National Standards Institute (ANSI). It has a very complex and comprehensive structure, only some of the relevant parts of which are shown in Figure 8.2.

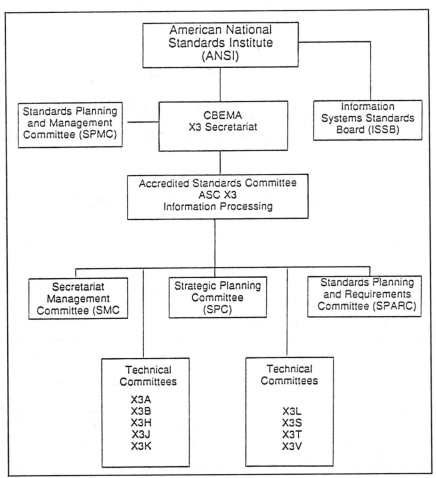

FIGURE 8.2   Structure of the American National Standards Institute (ANSI)

In order to put the U.S. standards-making process in perspective, a numerical breakdown of standards developed by each responsible agency as of 1992 is provided in Figure 8.3. The figure shows the various standards created in terms of a comprehensive pie-chart division.

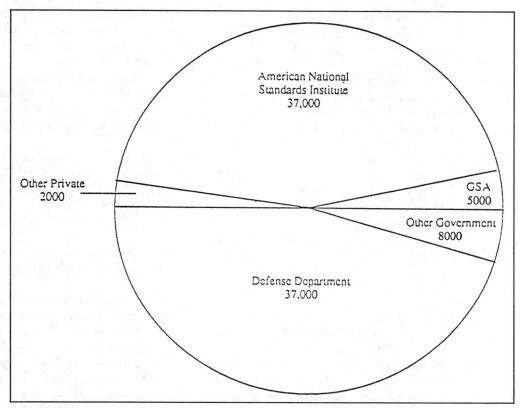

FIGURE 8.3   U.S. Standards Distribution

At one time the international entities such as the International Telecommunication Union, the International Standards Organization (ISO), the International Electro-Technical Commission (IEC), and the Institute of Electrical and Electronics Engineers (IEEE) tended to define standards from the top down, but today standards are more likely to be built from the bottom up at the national level or on the regional level in Europe. These "localized" standards are then somehow consolidated or "negotiated" at the top.

Although it is only a single case in point, the decision by the International Standards Organization and the International Electro-Technical Commission decision to declare October 14, 1994 as "World Standards Day" has turned out to have been a rather ironic occurrence. This is because, in fact, the celebration of this event turned out to be quite nonstandard. In practice, the United States celebrated the event on October 11, while Finland opted for October 13, and Italy straggled in on October 18th.[1]

Of the national standards-making systems, that of the United States is easily the most complex. The key entities involved in this process include:

(a)   The Exchange Carriers Standards Association (ECSA). (This group supports the T-1 Committee of the ANSI and all of its subcommittees.);

(b) The Electronic Industries Association (EIA) and the Telecommunications Industries Association (TIA). (These industry trade associations develop standards for data transmission, telephone terminals, optical communications, fax, and wireless services. They are responsible, for instance, for the key "IS" set of standards.);

(c) The North American ISDN Users Forum and the ATM Users Forum. (These groups, which geographically cover the United States and Canada, are coordinated by the National Institute of Standards and Technology (NIST) in the United States. These and other similar forum groups address standards and conformance testing across the boundary between user and vendors.);

(d) The American Federation of Information Processing Societies (AFIPS), as well as computer and digital telecommunications corporations that support the ANSI X.3 subcommittee on digital communications;

(e) The Corporation for Open Systems. (This group is a consortium of corporations dedicated to adopting viable "open standards" and is organized at both regional and national levels within the United States.);

(f) The Institute of Electrical and Electronics Engineers. (This professional association includes a communications society that addresses a number of telecommunications standards in such areas as wire and wireless LANs.); and

(g) The federal government. (In particular, this includes the Federal Communications Commission (FCC), the National Telecommunications and Information Administration (NTIA), the National Institute for Standards and Technology (NIST), the U.S. advisory committees for the ITU and its Consultative Committees, and the U.S. State Department. Other federal agencies such as the Occupational Safety and Health Agency (OSHA), the Department of Defense (DOD), and the National Science Foundation (NSF) can also play key roles in standards setting.)[2]

These types of U.S.-based standards-making groups are found in many other countries, but the U.S. structure contains by far the most organizations and participants. In short, at both the national and international level, the business of standards making has multiplied in complexity over the last two decades. The time needed to develop standards has, not surprisingly, tended to expand while the human resources required have spiraled upward in dramatic fashion.

These trends have created a variety of concerns. The developing countries have felt financially frozen out of the standards-making process because of the time, personnel, and expense that are involved. The commercial telecommunications industry has also been extremely critical of the slowness of the ITU standards-making process. This, in fact, has resulted in reforms started by ITU Secretary General Richard Butler over five years ago and implemented under the current ITU Secretary General Pekka Tarjannne. These reforms have set timetables and schedules for creating new standards that have shortened the process within the ITU. In the past, the development of new standards required from three to often as long as five years. The objective today is to create new standards on a timetable of two years to 18 months or even less.[3]

The type of standards that are created today can be of three basic types. A standard can, in fact, be an object such as a kilogram or a meter, or it can be "documentary" in terms of specified parameters and conditions, or it can even be conceptual. The documentary standard is by and large the most prevalent type. There are actually under the documentary standards three subcategories of basic standard, product standard, or an integrated systems standard. These types of documentary standards are set forth in Figure 8.4.

- Basic or Fundamental Standard
  — Establish basic principles for any industrial development
  — Examples: Units for measurement, precision of test methods
- Product Standard
  — Define Performance, etc., for product
  — Example: Tape Drive Standard
- Integrated Systems Standard
  — Define an Integrated System
  — Examples: OSI and ISDN

FIGURE 8.4   Type of Documentary Standards

The standards-making process operates at many different levels and with what might be considered an increasing level of formality. The type of standard that exists in the wireless telecommunications field starts with a company standard, which might be set for a product or a service. Next there is the industry-wide standard. After that come interindustry standards, then national and regional standards, and finally international standards.

The standards-making process is today perhaps best thought of as a multitiered process. This builds from industry and corporate initiatives and eventually becomes a global process. Many countries have a combination of governmental, professional society, and private industries or consortia that develop national standards. The U.S. structure as summarized above represents at least one example. There are often regional groups as well. In Europe, the most important units are the European Telecommunications Standards Institute, the Council on European Post and Telecommunications (CEPT), and the European Commission, especially Directorates XII and XIII. These units seek to coordinate and unify telecommunications-related standards for Europe. The power of a unified European position on telecommunications standards is considerable. The combined backing of dozens of countries, including many of the top ten economies of the world is sufficiently large that the European positions cannot be easily overlooked. In other parts of the world, regional approaches to standards are also common. In the Pacific region, there is a regional group known as the Asia-Pacific Telecommunity. This organizational unit, which was formed under the auspices of the ITU, considers regional plans for the Asia-Pacific region and coordinates standards issues for the organization's members.[4]

The same trend toward regionalism can be detected around the world. The strongest single force for regional standardization in the Asia-Pacific region is the so-called TTT Committee, which develops all national telecommunications standards for Japan. This highly sophisticated and powerful standards-making group for Japan also provides much of the leadership for the Asia-Pacific Telecommunity (APT).[5] In Europe, therefore, the ETSI and CEPT groups tend to amplify the power of their regional recommendations, and the APT tends to do the same for the Asia-Pacific region.

Nor are the Americas exempt from seeking some form of regional leverage. The Organization of American States (OAS) has a regional telecommunications group known as the

Committee on Telecommunications (CITEL), which addresses coordinated planning for the hemisphere, and especially telecommunications standards for the region. This is not to suggest that individual perspectives on telecommunications standards, frequency allocations, and planning do not develop within a region. Different groups within a region may strongly disagree. Nevertheless, regional solidarity often emerges behind key standards within Europe, within the Asia-Pacific Telecommunity, within the Americas, and sometimes within the African and Arab regional groups as well. Although sometimes the contentious issues involving telecommunications standards and frequency allocations are divided between the developed and developing countries of the world, it is today perhaps just as likely to see divisions and conflicting opinions along regional and trade groupings.[6]

Finally, it should be noted that at the global scale, telecommunications standards are not the sole preserve of the International Telecommunication Union. There are several other groups that have an important say in this process as well. These groups are the International Standards Organization, the International Electro-Technical Commission, and the Institute of Electrical and Electronics Engineers.

The International Standards Organization, headquartered in Geneva, Switzerland not far from ITU headquarters, is an important source of standards making in telecommunications. In recent years it has contributed to standards for ISDN, Open Systems Integration (OSI), and the seven-layer protocol under which OSI standards are being implemented.[7]

Another key international standards-making group is the International Electro-Technical Commission. This group, which represents manufacturers of electronics and telecommunications equipment, represents the formal process by which the private sector participates in international standards making.[8] Finally, there is the Institute of Electrical and Electronics Engineers (IEEE), which is a worldwide society of professionals in the field who help to develop and coordinate international standards in the field of electronics and telecommunications.[9]

Most international standards in the field of telecommunications, and especially those for wireless telecommunications, are developed within the ITU framework. The difficulty with wireless telecommunications, however, is that this is a new and rapidly evolving field and the provision of commercial services such as mobile cellular radio, digital cellular, and personal communication services (PCS) are initiated before the international standards are agreed upon. The decisions concerning the multiplexing technique to be used for PCS, and in particular, the choice between CDMA and TDMA is a matter of some dispute in the international standards-making arena.

This introduction is not intended to suggest that all standards-making efforts are overly cumbersome, bureaucratic, and regionally oriented. Many standards-making efforts are indeed highly productive and carried out in a cooperative manner. On the other hand, premature standards can stifle innovation and benefit only a few of the entrenched suppliers.

There is one recent trend that has been generally helpful to the business of creating standards in the field of telecommunication. In the past, the process of creating new telecommunications standards has been largely divided along technology-related lines and especially structured along the lines of creating either different or at least "two-tiered" standards for wire and wireless (or radio) technology. As digital communications tends to unite all telecommunications and information services and as hybrid networks tend to interlink cable and wireless systems, these old divisions have become increasingly passé. The ITU is being restructured to combine and integrate the standardization process, and regional groups can be expected to do the same in coming years.[10]

Most people involved in this reform and restructuring process believe that better and more integrated standards will be achieved and that the merger of cable and wireless technologies for standards-making purposes can also help to speed up the process as well.

8.2 THE KEY STANDARDS FOR WIRELESS COMMUNICATIONS SERVICES

The key standards and service categories for wireless communications services are described below in summary form. The brief technical description of the technical standard is accompanied by the service or services that are supported by that standard.

8.2.1 Mobile Wireless Terrestrial Services

There is only one analog mobile wireless standard, but in the digital area there will likely be a number of standards adopted around the world. The analog standard is known as Advanced Mobile Phone Service (AMPS) and it has replaced the precellular IMTS service throughout the United States.

Advanced Mobile Phone Service (AMPS)

This is the original cellular telephone standard, which has been in operation since the early 1980s. It is by far the most extensively deployed form of cellular wireless service in the world today. Despite the rather rudimentary nature of analog technology, it has proven versatile, and the number of AMPS subscribers in the United States has in the last decade gone from none to about 14 million and worldwide it has grown to nearly 26 million. There are well-defined standards for the creation of cells, MTSO operations and signaling, hand off of mobile service, billing procedures, and access charges.[11]

There are limitations with AMPS services that place it at a strategic disadvantage. One of these is the much higher capacity and potentially higher quality of digital systems and the other is the lack of reliable data transmission capabilities. There are limits to what can be done to improve the quality and throughput of AMPS systems beyond cell splitting and Narrow-AMPS enhancement techniques, but in the digital transmission area, cellular has developed a new standard for digital transmission. This new standard, known as Cellular Digital Packet Data (CDPD), creates a new carrier at the cell site to send high-quality data from remote mobile locations at rapid speeds that have superior performance over conventional mobile modems operating over a 3 kHz channel. This new cellular data carrier will allow this existing service to compete more effectively with ESMR and PCS, which are all-digital wireless services and thus already adept at providing data services. Finally, there is the upgraded AMPS standard for digital transmission using TDMA multiplexing. This is known as IS-55 and is also an EIA standard. There are about 1 million subscribers in the United States using this standard today.[12]

Global System for Mobile Services (GSM)

This standard and Advanced Mobile Phone Service (AMPS)/ IS-55 are the best defined and most extensively deployed of all mobile service standards on a global basis. GSM has been particularly widely accepted and implemented in Europe and the Asia-Pacific region of the world. Although other standards exist such as the U.K.'s DAC and the Scandinavian NMT-450 and NMT-900 these have not enjoyed any wide acceptance. GSM is a digital cellular service that employs the Time Division Multiple Access (TDMA) technique and is the first system to be fully constructed around the Signaling System Number 7 control protocol. This use of CCS Number 7 allows such value-added features as caller line identification, flow control, destination routing, and many ISDN features and connectionless services. The ISDN functionality also allows the set up of encryption messages, authentication of users at home and visiting locations, and dynamic allocation of channels. In its latest and most advanced form it is a fully interactive transmit and receive service that can be used in both vehicles as well as by pedestrian users. GSM has now been fully recognized by the JTC as one of accepted standards for PCS implementation in the United States.

The switching functions within a GSM network are handled by what is called a Mobile-Services Switching Center (MSC). The Center is the intelligence of the network and discharges the functions of routing, call-control, switching, plus all accounting and charging activities. Key elements of the MSC include the operations and maintenance center (OMC), the home location register (HLR), and the visitor location register (VLR). The referencing system is similar to that being defined for PCS services as provided in more detail in Appendix 3. The function is to allow users to be entirely mobile and to "roam" over a large geographic region and still obtain service, be properly billed, and still preserve anonymity.

In 1990, the some 17 countries, essentially all European, formally signed a memorandum of agreement agreeing to the application of GSM standards and to certify equipment built to GSM specifications. The GSM standards have now spread and are being accepted with North American and Japanese standards-making bodies.

This service was first developed in Europe under the auspices of the ETSI with strong industrial backing by the world's largest telecommunications equipment supplier, Alcatel; the original acronym of GSM stood for Group Speciale Mobile (GSM). Today, as the acceptance of the GSM standard has spread around the world, the more general phrase of Global System for Mobile has been substituted. Furthermore, the ETSI has also adapted to the more global focus by renaming its GSM Standards the Special Mobile Group.

The digitally based GSM system is based upon standards specified by the European Telecommunications Standards Institute and can be conceptually viewed as having three fundamental parts. These are the analog processing units that interface with the local network, the digital processing unit that is the core of the system, and the user interface unit. The most technically complex parts of this system are the digital processing unit and the speech processing within the codec equipment of the user interface unit.[13]

In the United States, digital cellular is offered under the standard known as IS-55 with nearly 1 million subscribers now using this service. The new PCS standards adopted by the Joint Technical Committee now encompasses GSM, IS-55, and three additional standards, which will be discussed in the following sections.

DECT and Telepoint (CT-1, CT-2, and CT-2+)

Digital European Cordless Telephone (DECT) is a so-called low-tier or pedestrian-based low-powered cordless telephone or narrow band PCS service. It is thus not designed for vehicular based mobile services, but rather for urban commercial use in high-density areas such as business centers, shopping malls, within office complexes, and so on. There is a fine line between these systems, wireless LANs, and wireless PABXs. These low-tier PCS services can be considered competitive with the Part 15 unlicensed IMS services, such as those offered by the Spectrallink Corporation. The licensed form of this lower-powered and pedestrian-based service is thus known as DECT.

A particular form of this type of digital wireless system that has been introduced on an unlicensed basis in the United Kingdom is known as Telepoint. The first version of Telepoint, known as CT-1, was a very rudimentary system and only allowed the receipt of calls. This required subscribers to go to a pay phone if they wished to call into the network. The latest rendition of the Telepoint standard, known as CT-2+, does in fact allow initiation of calls. Nevertheless, DECT and Telepoint are still so-called low-tier services in that they do not contain sufficient capability to handle vehicular traffic or other complex service functions. These are not likely to be sustained as primary standards for the future, but they do have the advantage of being very lightweight and low in cost.[14]

Personal Communications Service (PCS), Unlicensed PCS, and Wireless LANs

The problem with the PCS is that it is really more of a concept in the process of becoming a standard or a precisely defined service. In Europe the idea seems to be one of defining the next generation of Global Standard for Mobile (GSM) service as PCS and to define the next generation of Digital European Cordless Telephone (DECT) as the unlicensed PCS. In the United States, the thought process and certainly the decision-making process is much more complex. As many as 20 different options were first identified as possible PCS standards. This was subsequently narrowed down to seven possibilities. These were five standards for so-called high-mobility high-tier PCS services that could support pedestrian and vehicular traffic. In addition, there were two low-mobility or low-tier PCS services that would support cordless pedestrian-type services. These seven standards for PCS, in an extremely ecumenical move by the JTC, were all adopted as PCS standards for the United States. These are listed in Figure 8.5.[15]

High-Mobility Systems (high-tier):
- DCS 1800 (GSM)-Based (European TDMA digital cellular)
- IS-54-Based (US TDMA digital cellular)
- IS-95-Based (US CDMA digital cellular)
- W-CDMA (wide band CDMA)
- Hybrid TDMA/CDMA (Omnipoint Corp.)

Low-Mobility Systems (low-tier):
- DCTU [based on DECT (Digital European Cordless Telephone)]
- WACS/PHP based [mix of Bellcore WACS and Japan's Personal Handy Phone (PHP)]

FIGURE 8.5   PCS Technologies

In examining these seven standards, it is important to note several key factors. First, attempts have been made to group various possible technological solutions and to nominate the best "surviving" standard to represent that approach. The second key observation is that these standards have specific backers who have developed their own approach and that there is clearly an atmosphere of a "zero sum" game whereby if somebody else's standard wins you may very likely lose. This means that in this case and in many other cases of contemporary standards making, there is little sense of truly designing the "perfect" or ideal" standard that is best for everyone, but rather to "sell" standard that has already been developed by a particular group or organization. This situation can result in compromises (such as in the cell sizes) or to several standards that require conversion from one standard to another. This, in fact, is what resulted in this extremely complex standards-making process.

If one examines the seven standards for the new digital PCS standard for the United States, the key issues are actually quite transparent. There are really three key issues. These are well-proven TDMA versus not as well-proven CDMA technology, regional versus national trade consideration, and "low-tier" versus "high-tier" service standards.[16]

In the first case there is the proven, reliable, and broadly deployed TDMA technique (which has the worldwide acceptance of GSM behind it) versus the new, less proven but higher-capacity CMDA technique. In light of the acceptance of both techniques one might anticipate that systems will start with TDMA, but that the CDMA-based standard will become more popular in the 1990s, when the higher efficiencies of CDMA will be more important.

The second issue is that of U.S. national standards versus internationally developed standards. There is always a tendency to give some preference to a locally developed technology or standard that is considered homegrown. The broad acceptance of the GSM standard in Europe, Japan, and in other parts of the world suggests that this technology cannot be easily dismissed. Obviously, the first and second issues tend to become intermixed in such considerations.

Finally there is the issue of high-tier versus low-tier PCS service. Here, the JTC, when faced with a fork in the road decided to take it. Ultimately, the marketplace will decide about these services.

Now that PCS standards for the United States have been set in a most all-inclusive way operational services will likely be introduced in 1997 or 1998. The pacing item today is thus no longer standards, but rather freeing up the reallocated bands by migrating existing users to their new frequencies. Appendix 3 provides more detailed information as to how the PCS reference system will operate.[17]

One of the more critical components will be the provisions related to security. What seems very likely at this point is that there will be mandatory voice and data encryption. Further, the identification systems involving the international mobile subscriber identification, the temporary mobile subscriber identification, and so on, will probably be "highly encrypted" so as to provide PCS users with anonymity. This seems highly logical in that this is a benefit that PCS offers rather naturally as a digital service.

This feature of the PCS standard is actually a concern to the National Security Agency (NSA). This is because encrypted calls originating from a TAMS location make it virtually impossible to identify a mobile caller and to thereby monitor a particular targeted security risk. With 84 percent of the American public concerned to very concerned about privacy it seems likely that the encryption standard will be implemented despite the NSA's concerns. There is also another spinoff concern associated with the PCS security standards. The use of encryption technology currently controlled by U.S.-sensitive technology export regulations could be another issue in terms of trade. The manufacture and export of PCS equipment could become so restricted that U.S. manufacturers could be placed at a competitive disadvantage.[18]

There are a number of reasons why the standards for PCS have seemingly been slow to develop. Clearly there are a number of unresolved technical issues such as the right multiplexing technique (i.e., TDMA versus CDMA) the right power levels and cell or microcell size, and so on. One of the reasons for the delay, however, quite simply is the structuring of the standards-making groups that would be responsible for this task. After some deliberation it was agreed to create a so-called Joint Technical Committee on Wireless Access (JTC). This committee, which finally addressed and resolved U.S. national standards for 2-GHz PCS services, was a joint effort between the American National Standards Institute T1 Committee, in particular the T1P1-4 Committee and the Telecommunications Industry Association (TIA) Committee's TR 46.3.3 Committee. This group included federal governmental participants such as the National Telecommunications and Information Administration (NTIA), the FCC, the Defense Information Services Agency (DISA), and the National Security Agency (NSA); the various suppliers and vendors of PCS equipment such as Motorola, Zenith, NEC, Northern Telecom, and so on; and potential PCS operators.[19]

In addition to their efforts on high- and low-tier PCS they also had to undertake a parallel effort to develop standards for so-called unlicensed PCS as well. Again this was a highly sensitive issue in that this new unlicensed PCS will compete with existing Part 15 unlicensed IMS band. Ultimately, the JTC, beset by so many competing interests, technologies, and trade interests, simply opted to accept all seven of the finalists' standards.

Today there is an already established market in so-called wireless LAN and wireless PABX services. This is sometimes called "in-building wireless" to distinguish it from cellular. This service is largely a matter of business-based PABX service that is often overlaid on existing wire-based telephone services. It is less often used for a complete in-building telephone service or for data-based LAN services because it tends to be three to five times more expensive. It is, however, frequently used as an overlay system to extend capacity and to add mobility for those users who particularly need it.

Most of these currently available wireless services are operating in the frequency band from 902 to 928 MHz (nominally an Industrial, Scientific, and Medical (ISM) service, i.e., the FCC Part 15 (or unlicensed) band). There is some limited Part 15 service offering in the 49-MHz Land Mobile band, but this is constrained by the very narrow spectrum band available. There is also a 5-GHz Part 15 band that is designed for high-speed wireless ethernet-like services at 10 megabits per second. Although data can be provided through the wireless PABX systems, today these are circuit rather than packet switched, for under 1 megabit per second speeds and for the DECT standard discussed above. The spectrum can be used for in-building wireless LAN and PABX, but also for wireless PDA interconnects as well. Increasingly, however, PDA products such as Newton 2 are migrating to infrared bands.

The unlicensed frequency band has the advantage of not requiring any frequency coordination, but it also offers no protection. The only real standards requirement is that the product in terms of frequency bands, power levels, and operating systems must comply with the detailed FCC Part 15 rules. In the United States, wireless PABX and LAN standards under the DECT/CT-2+ standards will now be subsumed under the new wireless PCS standards. The availability of the unlicensed band for wireless PABX and LAN service offers smaller and more entrepreneurial firms a lower-barrier means to enter the market at a time when the auctioning of the PCS spectrum seems likely to be dominated by large and well-capitalized telecommunications carriers.

The development of these unlicensed wireless PCS standards was primarily undertaken by the TIA's Committee known as TR 41.6 and the WIN Forum. Finally, in the area of wireless LAN/PABX activities, the standards committee of the Institute of Electrical and Electronics Engineers (IEEE) has assisted in developing the new standards for the 2.4-GHz unlicensed band. This activity was carried out by the IEEE's 802.11 Committee.[20]

Specialized Mobile Radio (SMR) and Enhanced Specialized Mobile Radio (ESMR)

The standards issues in the area of terrestrial wireless are as complicated here as anywhere, simply because one service of longstanding and clear definition has been reengineered from within to become, in effect, a different service and a different standard. The key to all modern forms of wireless communications is, of course, access to favorable frequencies at reasonable cost and with extensive geographic coverage. The PCS auction in the United States has made a good deal of new frequencies available at 2 GHz and at 900 MHz but at very high cost. Further, obtaining a nationally uniform set of frequencies for seamless PCS service has proven to be enormously expensive.

The idea by some and most notably the NEXTEL system in the United States has thus been to seek to take over and redesign the outmoded single-channel analog SMR dispatching systems. In doing so they have developed new digital, cellular technology to address current SMR problems of low quality, contention for access, and lack of privacy. The new Enhanced SMR operators are converting the low-value and low-tech 14-MHz band to the latest in digital cellular technology. During the 1995–1996 time-frame many of the local SMR frequencies will have converted to EMSR systems and many others will likely be converted in the coming years.[21]

Universal Mobile Phone Service

This is a standards concept that has been generated within the CCIR group of the International Telecommunication Union. It has now been largely merged with the concepts now covered by Future Public Land Mobile Telecommunications Service, Universal Personal Telecommunication Services, or International Mobile Telecommunications 2000. The transition in the thought process reflected in these various names is important. The idea is that mobile telecommunications is no longer just a special service for a limited and highly targeted population. The idea is that it is becoming an integral part of a global telecommunications system geared to services to individuals rather than a location. (See the following section for more details.)[22]

Future Public Land Mobile Telecommunications Service (FPLMTS), Universal Personal Telecommunications Services (UPTS), and International Mobile Telecommunications 2000.

If AMPS is seen as the first generation of wireless mobile communications and IS-55, GSM, and PCS are seen as the second generation of wireless services, then FPLMTS or UPTS can be envisioned as the third and in some ways the definitive generation of wireless services. This concept is now also designated within the International Telecommunication Union as International Mobile Telecommunications 2000 or (IMT 2000).[23] The ITU Radio communications Study Group 8 has, for instance, sought to define FPLMTS as follows: "Future Public Land Mobile Telecommunications Services are third generation systems which aim to unify the diverse systems we see today into a seamless radio infrastructure capable of offering a wide range of services around the year 2000 in many different radio environments."[24]

Because this represents more of a longer-term planning process than a precise standards-making exercise, the basic concepts tend to be expressed in terms of goals. These goals include the following:

(a) an integrated mix of terrestrial fiber, coax, wireless, satellite, and other networks to provide both international roaming and truly global coverage anytime and anywhere;

(b) provide truly high-quality and high-throughput services to all subscribers, and in particular, to seek to establish ISDN objectives including digital service levels of 10^{-6} in terms of bit-error rates as well as megabit per second throughput.

The Future Public Land Mobile Telecommunications Service (FPLMTS) concept evolved from the wireless and radio communications sector of the ITU. Meanwhile, the Universal Personal Telecommunications Service (UPTS) evolved out of the international telecommunications or terrestrial "wire" sector of the ITU. They are now, in effect, merging together in the IMT 2000 concept. This makes a great deal of sense when it is realized that the communications systems of tomorrow will be digital and fully integrated in terms of service offerings. What is particularly powerful is the idea that to succeed UPTS/IMT 2000 must be fully integrated, seamless, and based on hybrid networks that combine cable, wireless, and satellite technologies.

This hybrid approach becomes almost inevitable when one accepts the idea of offering a telecommunications service to an individual rather than a place. Since people can be almost anywhere—at home, at the office, in a car or other mode of transportation, at a shopping mall, or a park—the idea of personal telecommunications service implies reaching someone wherever they may be. This means that wireless communications is a necessary part of the mix. It also means that rapid and efficient interconnection with fiber optic networks and terrestrial signaling and switching systems is also needed to create an efficient and cost-effective network.

What is especially noteworthy about the development of new standards in this area is that it is clearly being pulled by consumer demand and it is actually ahead of the technology rather than lagging behind it. Whether these standards are quickly achieved and the standards-making process stays ahead of the technology, however, remains to be seen.[25]

The complexity of the standards-making process for wireless telecommunications is so great and the rate of development of new standards so confusing, the chart shown in Figure 8.6 has been developed. In this chart, we thus attempt to show the evolution of basic analog IMTS services of the 1960s through to today's efforts to develop IMT 2000 and FPLMTS in the form of a flow chart. This is a conceptual presentation that actually simplifies the actual process, but it is still considered helpful in envisioning how the standards have generally evolved.

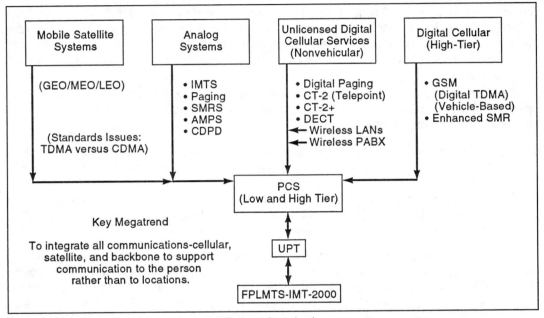

FIGURE 8.6   Key Historical Trends in Wireless Standards

8.3 STANDARDS FOR SATELLITE COMMUNICATIONS

The process of developing telecommunications standards has often consisted of setting the standards for wire-based telephone services and then seeing how they might need to be adapted or revised to other wireless or satellite systems. This has not surprisingly often tended to put satellite communications and wireless systems at a disadvantage. The Hypothetical Reference Connection (HRX) for telecommunications systems is based on a traditional wire-based telecommunications system with a hierarchical switched network architecture.

This means that this model assumes a reasonable degree of switching and concentration in the network. This means that the performance provides major allowances for switching and allocates only a small amount of the budget for network performance on the transmission path. In the satellite communications field, however, the reverse situation often holds.

The satellite link is almost all transmission and virtually no switching or processing. This may change in the future, but today the official noise allowance budget for satellite communications is largely out of "synch" with how it actually operates.[26]

The biggest single area of concern when it comes to satellite communications and standards is rather predictably in the area of transmission delay. Here the ITU has managed to develop a recommendation that allows for single-hop satellite communications links but does not accommodate double-hop connections. This "acceptable" standard for transmission delay of 400 milliseconds is complicated by today's networks. These systems may include not only a satellite link to geosynchronous orbit but also digital compression codecs, switches, speaker phone processors, and tandem terrestrial links. The cumulative effects of these various elements of delay can be latency effects that are considered unacceptable to users.[27]

As satellites move from being simply "bent pipes" to more complex systems with on-board processing and switching systems, the problem of latency can become even more pronounced. This probably will mean that geosynchronous systems will need very fast on-board processors or that low and medium earth orbit systems win out over higher orbit systems.[28]

The relative merits of different types of satellite systems and different orbital configurations were discussed in greater detail in Chapter 5.

8.4 OTHER WIRELESS STANDARDS ISSUES

The continuing evolution of wireless systems is creating not only new services and applications but also new wireless standards as well. These comprise, in many instances, a wireless form of existing wire-based services. Thus in many cases the existing standard is simply adapted or reapplied. The new types of wireless LANs are a case in point. The wireless LANs are, for instance, now subsumed under the overall IEEE-based LAN standards. The new wireless applications often present special issues of referencing, user identification, or special billing requirements when mobile services are involved. Other unique characteristics of the wireless technology also need to be addressed, but the existing standard can and often does serve as a major head start.

This building on the wire-standards base is an advantage to the standards-making process, but it also can put the wireless service at a disadvantage if special deviations or adaptations must be made. Wireless is thus still the exception rather than the norm. Also, the standards-making community is still predominantly terrestrial and wire based. The wireless standards-making community, a subset of the Telecommunications Industries Association, is often placed in an adversarial role vis-à-vis the other actors in the process. This evolution of a standards process that was based upon a division based upon a wire versus wireless split, even if not always formalized, can make the development of standards slower and more expensive at the national and the international levels.

In particular, the creation of Joint Technical Committees is increasingly dictated by the process. Although it seems as if this step is needed to reach broad agreement, it is clearly an expensive, labor-intensive, and at times highly inefficient process. In time an integrated approach to hybrid networks that develops neutral integrated and technology standards may evolve, but for the present, the evolution of wireless standards from their wire-based precursors will likely continue. Institutional reforms within the ITU to create a unified standards-making unit to replace the CCITT and CCIR is a step forward. Likewise the move to create integrated standards for Universal Personal Telecommunications Service standards is another clear step in the right direction. Perhaps, in the next few years parallel trends at the national level will also emerge. This new structure is reflected in Figure 8.7.[29]

```
1) The Plenipotentiary Conference
2) The Council
3) The Radiocommunication Sector
     a. world and regional radiocommunication conferences
     b. a Radio Regulation Board
     c. radiocommunication assemblies
     d. radiocommunication study groups
     e. a Radiocommunication Bureau headed by an elected director
4) The Telecommunication Standardization Sector
     a. world telecommunication standardization sector
     b. telecommunication standardization study groups
     c. a Telecommunication Standardization Bureau headed by an elected director
5) The Telecommunication Development Sector
     a. world and regional telecommunication development conferences
     b. telecommunication development study groups
     c. a Telecommunication Development Bureau headed by an elected director
6) The General Secretariat
     a. 600 employees
     b. an elected Secretary-General
```

FIGURE 8.7   Structure of the new ITU

8.5 SUMMARY AND CONCLUSIONS

The importance of standards to the development of wireless technology is not to be underestimated. The resolution of the issues related to the new digital standards for mobile cellular service is currently the crucial element in the future evolution of this field. There are, in fact, several key areas where standards will play a key role. These areas include the following.

Converging Technologies and Market Sectors.   Standards are often developed on a specific technology and service basis, but in today's world of convergence, the crossover effect becomes a complicating factor. The ESMR, IS-55, GMS, and PCS services are today separated by standards but united by converging markets.

Synoptic Families of Standards.   Increasingly there is an impetus to create a range of integrated standards such as ISDN and broad band ISDN. Under this comprehensive family of standards, the levels or objectives are set for quality, system availability, throughput rates, overhead levels, and so on. It is possible that these broadly integrated standards can allow a more efficient and seamless merging of wire and wireless services based upon hybrid networks.

Rapid Technical Evolution in the Wireless Telecommunications Sector and the Need for Dynamic Standards. The rapid development of wireless technology can only create tension within the standards-making process. The key issues include the following: (i) changing shape and characteristics of microcells for PCS; (ii) the new digital multiplexing techniques that allow more frequent reuse techniques within cellular; and (iii) the increased opportunities for patterns of frequency reuse and improved encoding techniques. These factors and more tend to redefine the whole concept of standards making. Instead of a single process, it becomes an ongoing iterative process, like the issuance of new versions of software.

Institutional Changes in the Standards-Making Process. The world that divided the standards-making process into parts labeled "wire," "wireless," and "satellite" is becoming obsolete. New institutional structures are needed to address hybrid and integrated systems as if they are seamless systems. This overall approach needs to include the "entire system," such as user terminal equipment that processes data and introduces delay into the transmission system even before it is a part of the "formally defined telecommunications network."

Today, the process of defining telecommunications standards is itself a billion dollar enterprise. It has a major impact on an industry that will soon be a trillion dollar undertaking. Furthermore, as the divide between the computer and information science industries becomes more and more fuzzy, the standards process for electronics, optoelectronics, and telecommunications will increasingly merge. Within a decade, there will be one vast $3 trillion information and telecommunications market and the standards-making process, for better or worse, will need to adapt. As all optical and electronics equipment for every conceivable application become "software defined," this need for integrated standards making will have become of vital importance.

ENDNOTES

(1) "Cable and Standards," *SPECS*, September 1992, Vol. 4, No. 7, pp. 2-3; also see *Open Systems Today*, October 30, 1994, pp. 2-5.

(2) *Open Systems Today*, October 30, 1994, p. 8.

(3) Codding, G., "Reforms in the ITU Standards-Making Process," Lecture, University of Colorado at Boulder, April 1994, pp. 1-18.

(4) Feld, W.J., and R.S. Jordan, *International Organizations: A Comparative Approach* (New York: Praeger Special Studies, 1983), pp. 201-212.

(5) ——, *International Organizations: A Comparative Approach* (New York: Praeger Special Studies, 1983) pp. 213-215.

(6) ——, *International Organizations: A Comparative Approach* (New York: Praeger Special Studies, 1983) p. 220.

(7) Stalling, W.J., *Handbook of Computer-Communications Standards* (Carmel, IN: H.W. Sams, 1989), Vol. 1, pp. 27-36.

(8) ——, *Handbook of Computer-Communications Standards* (Carmel, IN: H.W. Sams, 1989), Vol. 1, pp. 70-81.

(9) ——, *Handbook of Computer-Communications Standards* (Carmel, IN: H.W. Sams, 1989), Vol. 1, pp. 120-141.

(10) Reed, K., *Reorganization of the CCITT and the ITU* (Boston, MA: Arlex, 1991), pp. 102-105.

(11) Calhoun, G., *Digital Cellular Radio* (Norwood, MA: Artech House, 1987), pp. 103-106.

(12) DeSilva, E.W., "PCS/Wireless Regulatory Update" (ICA Exposition, Dallas, TX, May 24, 1994); also see *Newton's Telecom Dictionary* (New York: Telecom Library Inc., 1994), p. 489.

(13) Beaudry, M., and G. Parker, "Global System for Mobile Communications," *Telesis*, July 1992, p. 55ff; also see Ahola, K., "Europe's GSM: Passage to Digital," *Telephone Engineer and Management*, September 15, 1992, p. 59ff.

(14) Holland, B., "In-Building Wireless and PCS Services" (Boulder, CO: University of Colorado, September 1994), pp. 1-25.

(15) Laflin, M.G., "Personal Communications in a Changing World," Institute for Telecommunications Sciences, U.S. Department of Commerce, Technical Report, Washington, DC, June 1994, pp. 12-15.

(16) ——, "Personal Communications in a Changing World," Institute for Telecommunications Sciences, U.S. Department of Commerce, Technical Report, Washington, DC, June 1994, pp. 16-17.

(17) ——, "Personal Communications in a Changing World," Institute for Telecommunications Sciences, U.S. Department of Commerce, Technical Report, Washington, DC, June 1994, pp. 22-23.

(18) Branscomb, A., *Who Owns Information* (New York: HarperCollins, 1994).

(19) Laflin, M.G., "Personal Communications in a Changing World," Institute for Telecommunications Sciences, U.S. Department of Commerce, Technical Report, Washington, DC, June 1994, pp. 24-25

(20) Holland, B., "In-Building Wireless and PCS Services" (Boulder, CO: University of Colorado, September 1994), pp. 4-12.

(21) Schnee, V., "NEXTEL's ESMR Arrives: A New Era of Cellular Integrated Services," *Wireless*, January/February 1994, pp. 17-21.

(22) Allen, K., and M.G. Laflin, "UPTS, IMT 2000, and FPLMTS," Technical Report, U.S. Department of Commerce, May 1994, pp. 2-10.

(23) ——, "UPTS, IMT 2000, and FPLMTS," Technical Report, U.S. Department of Commerce, May 1994, pp. 22-27.

(24) Report of Task Group 8/1, ITU Radio Communications Study Group 8, April 1994.

(25) Allen, K., and M.G. Laflin, "UPTS, IMT 2000, and FPLMTS," Technical Report, U.S. Department of Commerce, May 1994, pp. 11-13.

(26) Lewis, G., *Communications Services Via Satellite*, 2nd ed. (Oxford, U. K.: Butterworth-Hiennemann Publishers, 1988), pp. 63-66.

(27) Morgan, W., *Communications Satellite Handbook* (Boston, MA: Wiley Interscience, 1990), pp. 101-122.

(28) Pelton, J.N., *How To Book of Satellite Communications* (Sonoma, CA: Design Publishers, 1992), pp. 71-79.

(29) Reed, K., *Reorganization of the CCITT and the ITU* (Boston, MA: Arlex, 1991), pp. 106-108; also see Codding, G., "Reforms in the ITU Standards-Making Process," Lecture, University of Colorado at Boulder, April 1994, pp. 20-24.

PART IV

EXPLORING THE FUTURE AND CONCLUSIONS

CHAPTER NINE

THE FUTURE OF WIRELESS AND SATELLITE TELECOMMUNICATIONS

9.1 INTRODUCTION

There will clearly be a future of wireless telecommunications as long as there are people on the move. Wire and cable communications will provide the predominant share of the communications backbone infrastructure of the twenty-first century. In terms of throughput capacity, quality of service, low latency and cost per bit of information transmitted, fiber optic cable will be difficult to exceed in terms of raw technical and financial performance. Wireless service will, however, continue to develop in its areas of strength. These are convenience, flexibility, mobility, and immediate accessibility. These are sectors where the general consumer and business people will likely place increasing importance and value in coming decades. Consumers have proven they will pay four to five times basic telecommunications rates for immediate connectivity and mobility.

Furthermore, broadcasting services and rural and remote access will also likely appreciate in value. The consistent evidence presented throughout this text suggests that the only logical projection for the future is that there will be integrated, seamless, and hopefully, well-designed "hybrid" networks that use both air and glass for instantaneous connections. The future of communications is seemingly in Universal Personal Telecommunications Service (UPTS), which will combine wire and wireless communications on the ground as well as in space and even in so-called proto-space (e.g., the new in-between environment inhabited by High Altitude Long Endurance (HALE) communications platforms. The challenge for wireless telecommunications in the twenty-first century may well bear a number of similarities to those encountered today. There will likely still be concerns about compatible standards, about new and broad band frequency allocations, and about scarce frequency resources. These will be coupled with concerns about the increasing cost of spectra in an environment where there are auctions of reallocated frequencies to the highest bidder. The stresses of the strains of national versus regional versus international auctioning of frequencies may well become one of the key dilemmas for global telecommunications officials to solve in the next few years.

Certainly, there will also be concerns about how wireless systems can produce broad band throughput at lower costs. Spiraling costs of frequencies through auction processes will tend to make that increasingly difficult.

9.2 NEW TRENDS FOR THE FUTURE

The remarkable growth and development of wireless during the last part of the twentieth century seems very likely to continue into the twenty-first century, but clearly there will be some key changes. Some suggestions have been made that fundamental shifts in the system for allocating frequencies could be made. One proposal has come from telecommunications guru George Gilder, who has suggested that the old system of frequency allocation coupled to high-priced frequency auctions represents the "exact" wrong way to proceed. Gilder has proposed the implementation of new technology to eliminate the need for the licensing and allocation of frequencies at all. He has advocated the use of techniques such as spread spectrum (or CDMA) but operating across very wide dynamic ranges of frequencies that spans large bands of the spectra.

Frequencies that are available at a given instant would be selected and used. Gilder argues that traditional frequency allocation processes could become obsolete. In the new environment envisioned by Gilder, allocations would simply not be necessary. A scanner would simply search through the spectra for an available frequency across a very wide dynamic range and use whatever frequencies were available. The ability to search and assign frequencies on demand across many different options would serve to create the impression of an almost infinite spectra being available for any and all applications.[1]

The logic of such a dynamic and on-demand temporary frequency allocation system may indeed lead to its implementation at some future point, but today's institutional "ownership" of bands for different telecommunications and broadcasting purposes gives a great deal of weight and "sunk investment" inertia to the status quo. The recent multibillion dollar spectrum auction for personal communications services in the United States and the worldwide process of assigning licenses for radio and television stations all serve to undercut the possibility of near-term change in this area.[2] "Here, as in so many other aspects of telecommunications, the difficulty of changing from an old to a new technology or from one institutional pattern to another is very hard to accomplish without a very long transitional period. This sense of inertia blocking change and innovation recalls the ironic observation of a telecommunications network planner: 'God created the earth in only seven days, but then again there was no imbedded plant.'"

There are a number of changes that seem very likely to occur in the not-so-distant future, however, that can be usefully discussed and assessed.

Higher Cost for Frequencies.   Rapid fluctuations in the process for allocation, assignment, and charging for frequency spectra has created apprehension and market fears that have recently served to create pressures to value available band-widths at higher and higher values. While new technology will be developed to achieve more effective use of available frequencies (i.e., spread spectrum, digital compression techniques, microcellular reuse systems, and dynamic band selection over very large frequency ranges), there will also be parallel efforts to charge more and more for access to the spectral. This is simply a matter of increasing market demand.[3]

Broader Band Mobile Services.   In parallel, there will be increased demand for broad band applications. As high-quality voice and data services become available in mobile wireless systems, the next step is clearly to seek a means to provide broader band services such as video, imaging, and higher-speed data in mobile systems and to do so at significantly lower costs.[4]

Increasingly Hybrid Systems.   The logic of maximizing the mutual capabilities of wire and wireless technologies is at least implicit in the two previously noted trends. There is considerable merit to the idea of using fiber optic cables along highways, in office building riser systems, and within other high-density areas. These fibers can then be linked to wireless cellular, LAN, or PABX systems. This is to suggest that wireless systems can and should be optimized for broadcast and mobile network usage and rural and remote areas. Fiber should then be used for concentrated high-throughput systems. This would allow wireless systems to interconnect at the nearest available and cost-effective interface. Cable TV systems could be linked to wireless communications networks to serve the information and entertainment needs of the family throughout the house, the outdoors, the backyard, or immediate neighborhood. This formula is not to suggest that satellites and wireless cannot or should not provide broad band services. It only suggests that these should be wireless, selected and oriented toward broadcasting services.[5]

Software is the Key to the Future.   The next few years will see software becoming more and more dominant in defining networks and enhancing value. The new digital environment will perhaps see such innovations as "software-defined radios," "software-defined television sets," and "software-defined cellular transceivers." This is to say that with the addition of the right software to a portable computer and you could "create," simply with the right new instructions, almost any electronic device you want from an oscilloscope to a video monitor. This suggests that wireless systems of the future could likely be reconfigured and updated more quickly and that modular systems with software-defined performance characteristics could be substituted quickly, efficiently, and at low cost. Perhaps even more importantly, systems hardware in the future could be upgraded or retrofitted simply by adding a new version of software.[6]

The Power of Digital.   The gains achieved by digital technology in the wireless communications sector have been extremely impressive with an almost tenfold gain in efficiency over the last decade. The gains have come in terms of improved modulation and multiplexing techniques, new encoding schemes, and improved compression algorithms. There is good reason to believe that similar types of gains will be achieved yet again over the next decade and that this can be achieved without violating the theoretical bounds of Shannon's Law.[7]

Movement Up the Electromagnetic Spectra.   The last century has seen a persistent movement toward the use of higher and higher frequencies to respond to the needs of more and more users desiring more services involving broader band telecommunications requirements. There seems to be a high degree of likelihood that this will continue with upward movement to higher and higher frequencies. This will include millimeter wave and infrared and even free-space optical communications being used for wireless applications. Infrared communications as shown by television remote controls and personal computers can provide very effective short-range communications with high reliability. Wireless LANs, PABXs, and other transmission and switching systems will certainly follow suit. Firms such as Motorola, Spectrallink, and others have identified this as a key new market for the next decade and beyond. Although the movement up the spectra for wireless services may tend to be focused toward localized service, it is possible that high-energy systems in the ultraviolet up to the cosmic wave regions might be adapted to such exotic concepts as interplanetary communications.[8]

Flexible and Innovative Standards.   The current morass of standards making for wireless mobile telecommunications is seemingly moving toward resolution. The so-called third generation of mobile services, which revolve around the prospective standards known as FPLMTS, IMT 2000, and UPT, seem intent on accomplishing several key objectives. These objectives include a "true" global standard, an integration of wire and wireless networks to achieve a seamless network, global roaming, a universal personal telephone number, and a new concept of communications based on calling people rather than locations. If implemented as now conceived, this new UPT standard will serve to integrate all telecommunications into seamless systems that can optimize the best of wire, coaxial cable, fiber optic, satellite, HALE, and the various terrestrial-based wireless technologies.[9]

9.3 NEW BREAKTHROUGHS FOR THE TWENTY-FIRST CENTURY

It will perhaps seem that the above-projected trends for the future are essentially extrapolations of current trends and as such are rather predictable. In fact, the above are essentially all incremental forecasts. Even so, "predictable forecasts" are useful if they indeed come true. Some may believe even these "extrapolated projections" are on the optimistic side. When one considers the exponential gains in telecommunications over the past decade, however, these predictions do not seem overly extreme. The past decade strongly suggests that even more than a tenfold gain could be achieved in terms of efficiency, throughput, and cost performance within the next ten years. This is essentially the Moore's Law prediction, which has held true for computers for 30 years.

The emerging science of complexity theory suggests that future developments are often nonlinear and that innovations can unexpectedly occur. Some of the possible changes, breakthroughs, or technological inventions in wireless telecommunications that might be characterized as "nonlinear" developments are as follows.[10]

The "Smart" Environment. The application of telecommunications capabilities and processing power to almost everything is certainly a plausible next step. The even-lower cost of processors and the extension of that processing power to virtually every fixed and mobile location via wireless telecommunications technology is becoming increasingly feasible. Intelligence in buildings will allow us to call from hallways and elevators. The energy use in our homes could be automatically controlled, while television broadcasts and electronic newspapers can be easily sent to cars, buses, trains, or airplanes by the early twenty-first century. Perhaps the rarest commodity in the decades ahead will be a totally media- and technology-free environment.

In theory, the new "smart" environment will allow us to have quicker fire, police, medical, or other emergency services. It will also afford us a wider range and potentially more cost-effective range of entertainment, information, library, education, and governmental services not only in the home or office, but anywhere we might be. The certainty seems to be that our environment will be not only smarter but immediately accessible via a range of wireless technologies that will include not only radio signals but infrared and maybe even free-path optical systems.[11]

One generic illustration of how "intelligent" structures can integrate systems and subsystems is presented in Figure 9.1. It presents a schematic of how distributed floor-control, facilities management, and operational systems might be combined in a future building environment.[12]

Tele-Computer-Energetics. The convergence of telecommunications, computers, and now even energy systems is one of the key potential megatrends of the twenty-first century. The convergence trends that are today linking communications, cable television, consumer electronics, computers, and content (e.g., newspapers, electronic databases, software) will soon also encompass transportation and energy systems. Perhaps equally important, all of the systems appear to be moving toward flexibility and mobility. This can be expressed in terms of innovative unitized and regenerative fuel cells for electric cars, computers, or cellular telephones, or it can be expressed in terms of satellites or HALE platforms that can transfer energy from one location to another.

The dual phenomena of convergence and energy crossover into the computer and telecommunications fields will likely serve to drive the cost of wireless devices downward. It will also likely extend the breadth and depth of the applications across this very broad range of activities covered by the admittedly difficult phrase "tele-computer-energetics." This will likely involve not only the intersection of many technologies but also will help us better engineer the "smart" environment described above.

These new applications of tele-commuter-energetics may involve areas such as car and truck navigation, enhanced safety and rescue operations, advanced new construction techniques, electronic newspaper distribution, provision of power and information to rural and remote areas of the world, renewable energy systems for transportation, and bioenergy systems. In short, one of the key forces that is driving us toward the so-called "smart" environment comes from the three-way convergence inherent in the tele-computer-energetics phenomena.[13]

The Ultra-Personal Digital Assistant. Today the concepts variously known as the personal digital assistant, intelligent agent, and personal communicator are allowing com-

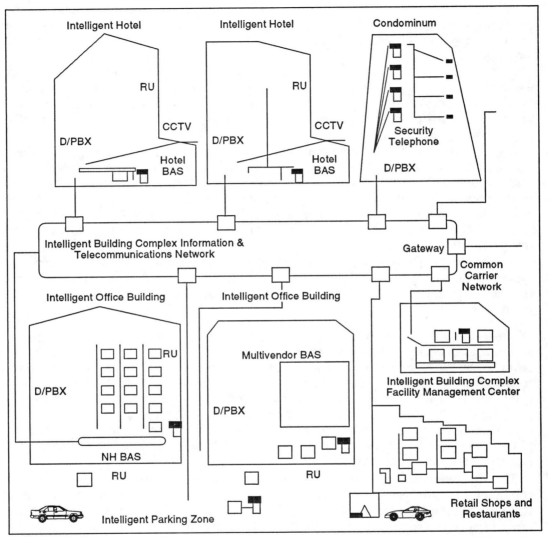

FIGURE 9.1 Intelligent Building Complex

plete mobility for the executive on the move. These small, compact, "intelligent," and highly versatile devices have helped to redefine the capabilities covered by the phrase "virtual office." The integration of such functions as the cellular telephone, mobile fax, personal computer, mobile e-mail, personal data assistant, and even the portable radio and television receiver is being driven by software-defined electronic devices.

This powerful trend seems to be moving toward more powerful intelligence, faster processing times, and broader band capabilities at lower costs and in more compact formats. This almost suggests that the ultimate personal digital assistant might be like the Star Trek communicator, where voice instruction can control every possible function from data entry up to and including broad band imaging.

This idea of an extremely portable device that can provide an enormous amount of information and telecommunications services is something of a technological marvel, but it also could be seen as a threat. This "ultimate intellectual prosthetic" that can almost think for us is in some ways a very intimidating device. If such a unit can filter information, prioritize messages, prepare responses to inquiries, and arrange meetings, one might ask what there is left for a human to do. An illustration of this type of intelligent agent/personal digital assistant is provided in Figure 9.2. It is anticipated that the sophistication levels will exceed the early systems in only a short period of time.[14]

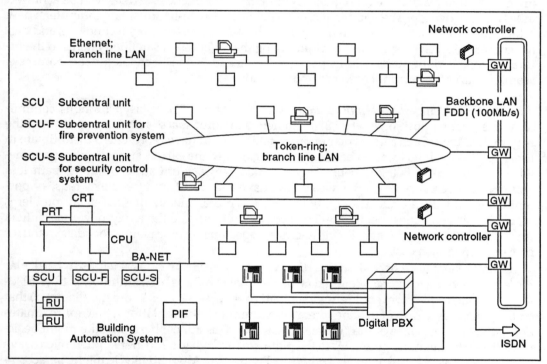

FIGURE 9.2 Designing Intelligent Structures and Environments

The Satellite Systems of the Twenty-First Century. The continuing rapid development of fiber optic systems with a current build rate of almost one mile per minute during the business day is certainly prodigious. Further, the rapid build-out of cellular radio systems, and soon, PCS networks may well also serve to stimulate a parallel and very rapid growth in the wireless environment. The opportunity for satellite communications to supplement and extend these rapidly expanding fiber and cellular networks is about to increase sharply as well. This is due to several factors. The advent of lower earth orbits can reduce the problems of latency and echo and can also expand the opportunity for frequency reuse. Further, high-speed on-board satellite processing can promote broad band operation at increasingly more cost-effective levels. Finally, the advent of intersatellite links in LEO, MEO, and GEO orbits will enhance connectivity, allow more direct routing, and create improved networking efficiencies.

New high-powered and digitally compressed Direct Broadcast Satellite services can supplement cable television services in terms of broad band communications to the home. In short, in the field of satellite communications, there has been more new satellite technology in the last two years than in the decade that preceded it. There is a clear potential for satellite communications to provide broad band, flexible, mobile, and low-latency services that are indeed competitive with fiber optic systems at least at certain levels and for certain applications. The potential of low and medium earth orbit satellites to create a whole new image for satellites that are high quality with virtually no echo and delay can be achieved in less than a decade. The combination of satellites with on-board super computers, new orbital configurations, new frequency allocations, active phased array technology, and digital compression techniques are revolutionizing the potential of satellite systems in the future. With the right standards in place, satellite and HALE technology can become a seamless part of terrestrial-based communications systems.[15]

The High Altitude Long Endurance (HALE) Platform.   The geosynchronous communications satellite positioned at 35,870 kilometers in outer space is almost one-tenth of the way to the Moon. Such a geosynchronous satellite would have represented the ultimate in useful platforms if its location had been much closer to the earth's surface, perhaps somewhere in the 25- to 800-kilometers-high altitude range. From such locations, path loss would have been very low, coverages would have been ideal, and frequency reuse opportunities would have been truly excellent. Furthermore, latency or time-delay problems would have been almost negligible. The geosynchronous satellite today, of course, suffers from problems of path loss, inefficiencies in spectra reuse, transmission delay, and of course, high launch costs.[16]

Recently there have been thoughts of a "virtual" geosynchronous satellite. In particular there has been the idea of an "eternal airplane" that can fly at high altitudes of 18 to 20 kilometers. Such a platform could provide broad coverages (i.e., 500 kilometer diameter) that could be used as a form of a flying interactive communications platform that approximates at least the performance of the "ultimate" satellite. This atmospheric satellite might be accomplished by beaming power up to the platform through ground-transmitted microwave or millimeter wave antennas, as shown in Figure 9.3. Alternatively it might be accomplished by using high-altitude reciprocating propeller systems or solar-cell power with fuel-cell storage. This type of HALE platform operating in so-called proto-space might look and act more like a very-high-altitude aircraft as shown in Figures 9.4 and 9.5. The advantage of this second approach is that the platform can remain stationary or it can flexibly move to other locations for observation, remote sensing, surveillance, or other such mobility-driven applications.[17]

Although such platforms might be used for a number of purposes, the most viable application would almost undoubtedly be wireless telecommunications for television, radio, cellular telephone, disaster warning and recovery, and even rural telecommunications, education and medical services. The power for maintaining the platform in orbit can come from several sources. These include photo-voltaic cells, regenerative, unitized fuel cells,

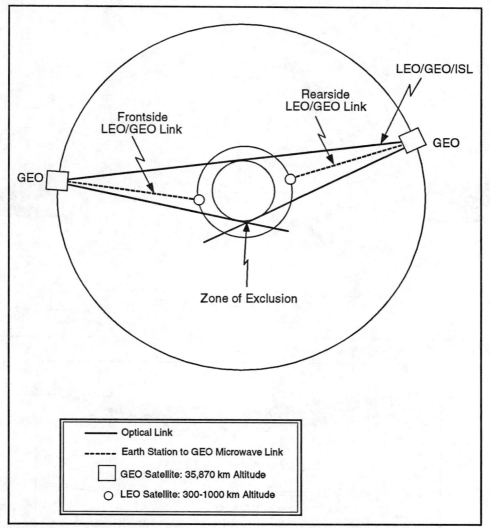

FIGURE 9.3 Intersatellite Links: Geosynchronous to Geosynchronous and Low Earth Orbit to Geosynchronous LInks

and even microwave or millimeter wave power transmitted from the ground up to the platform.

The birth of several corporations to support the commercial development of such HALO platforms could have a major impact on the world of telecommunications. These new platforms could provide extremely good and cost-effective services to developing countries. They could also be used as an add-on adjunct to some of the new low earth orbit satellite systems such as Iridium and Teledesic. In general, the ability to develop new types of platforms to fly in "proto-space," between aviation space and outer space, could represent a

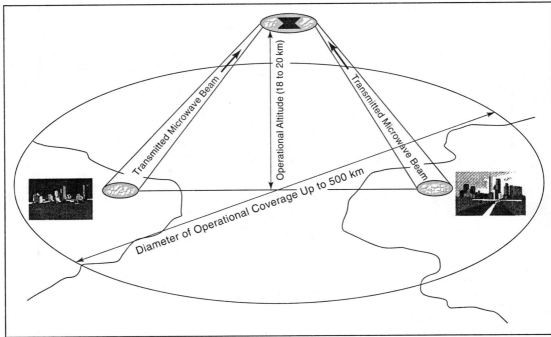

FIGURE 9.4   High Altitude Long Endurance (HALE) Platform

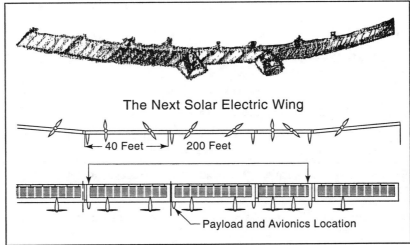

FIGURE 9.5   High Altitude Long Endurance Platform

new departure for the field of wireless telecommunications, especially in the developing world.[15]

It should be further noted that the deployment of HALE platforms in proto-space, especially if directly connected to broad band satellites, could create a host of new capabilities. These new applications might include earth observation for agricultural, forestry, or ocean monitoring purposes, electronic news gathering, crime and fire detection, and even electrical power relay. These platforms could operate as very stationary "virtual" towers, or they could be designed to "fly" at very high altitudes to, say, cover a news story or observe a forest fire or an oil spill.

9.4 SUMMARY AND CONCLUSIONS

The rate of innovation in the field of wireless telecommunications can be expected to remain rapid and perhaps more than a little unpredictable. The basic patterns of change would seem likely to include: (a) increasing use of higher and higher frequencies, including employment of infrared and optical systems for free-space communications; (b) higher costs and greater scarcity of spectra; (c) evolution of new standards that promote seamless hybrid systems that link wire and wireless networks together; (d) increasing use of more advanced digital modulation, encoding, multiplexing, and compression techniques; (e) increasing development of complex, small, and lower-cost mobile devices that can perform an entire range of "virtual" office and intelligent agent functions; (f) software-defined equipment (this means that software will define radios to computers and cellular telephones to television sets); and (g) a staggering range of new technologies that cut across every sector of societal, cultural, economic, and political life.

These new technological developments include such aspects as:

- the "smart" home, building, and highway;
- the tele-computer-energetics technological convergence;
- the ultra-personal digital assistants;
- advanced regenerative and processing satellites with on-board super computers;
- high altitude long endurance (HALE) platforms;
- mobile telecommunications devices that merge virtually all office and computer functions and capabilities together into a single integrated system that is both portable and low cost. (These may allow new education and health applications like the so-called Electronic Tutor.)

These new developments in wireless technology and services are intended only to be descriptive of the future rather than complete or exhaustive. If the market and consumer enthusiasm for portability, mobility, flexible accessibility, and rapid reconfiguration of telecommunications resources continues into the twenty-first century unabated then these products and services can be counted on to succeed. The realization of elaborate, cost-effective, and seamlessly integrated "hybrid" wire and wireless networks may represent one of the major accomplishments of the next century. Such systems could help to create a new human capability—the ability to think and interact instantly and simultaneously with anyone on the planet anytime and anywhere.

ENDNOTES

(1) Gilder, G., "Telecosm: Auctioning the Airways," *Forbes*, April 1994, pp. 98-112.

(2) "Frequency Auction Exceeds Forecasts," *Miami Herald*, July 23, 1993, p. B1.

(3) "Frequency Auction Exceeds Forecasts," *Miami Herald*, July 23, 1993, pp. B5-B7.

(4) Channing, I., "Customers Wanted: Prospects for Personal Communications Services," *Communications International*, September 1993; also see Channing, I., and R. Burr, "Worldwide Digital Cellular," *Mobile Communications International*, Winter, 1994.

(5) Frieden, R., "Wire vs. Wireline: Can Network Parity Be Reached?" *Satellite Communications Magazine*, July 1994, pp. 20-23.

(6) Gilder, G., "Telecosm: Auctioning the Airways," *Forbes*, April 1994, pp. 37-40.

(7) Calhoun, G., *Digital Mobile Communication* (Norwood, MA: Artech, 1990), p. 17ff.

(8) Manuta, L., "Riding the Spectrum Wave?" *Satellite Communications*, July 1994, pp. 24-25.

(9) Hoffmeyer, J.A., "Personal Communications Services," Institute for Telecommunications Sciences, Staff Paper on PCS Standards, Boulder, CO, July 1994, pp. 1-20.

(10) Waldorf, M., *Complexity: The Emerging New Science at the Boundary of Order and Chaos* (New York: Touchstone, 1992), pp. 198-222.

(11) "Future Topics: Current Trends in Intelligent Buildings and Environments," *IEEE Communications Magazine*, October 10, 1993, Vol. 31, No. 10, pp. 67-70.

(12) "Future Topics: Current Trends in Intelligent Buildings and Environments," *IEEE Communications Magazine*, October 10, 1993, Vol. 31, No. 10, pp. 72-73.

(13) Pelton, J.N., *Future View: Communications, Technology and Society* (Boulder, CO: Baylin Publications, 1992).

(14) Bernstein, P., "Intelligent Agents: It's Magic," *Wireless*, March/April 1994, Vol. 3, No. 2, pp. 18-21.

(15) Shimamoto, N., "Communications Satellite Technology," *Via Satellite*, July 1994, pp. 18-22.

(16) Glaser, P., "The Power Relay Satellite and HALE Platforms" 44th Congress International Astronautical Federation, Plenary Lecture, October 16-22, 1993 Graz, Austria, p. 2.

(17) ———, "The Power Relay Satellite and HALE Platforms" 44th Congress International Astronautical Federation, Plenary Lecture, October 16-22, 1993 Graz, Austria, pp. 1-16.

CHAPTER TEN

CONCLUSION

10.1 INTRODUCTION

The complexity of wireless telecommunications seem undeniable. There are in excess of 30 clearly defined wireless services being provided worldwide and most of these are still being provided in the United States. Some of these, such as tropo-scatter, short-wave radio communications, and even microwave relay, are being phased out of operation. Other new and innovative services such as personal communications service (PCS), Universal Personal Telecommunications (UPT), personal digital assistants (PDAs), low and medium earth orbit mobile satellite communications, and direct broadcast satellite services are in various stages of being phased into service.

Some of these services are outside the "normal" commercial wireless marketplace in that they represent military services, or federal, state, or local governmental services. These include public safety, search and rescue, or other governmental services. Finally there are other "special" wireless communications activities such as amateur radio, environmental monitoring, humanitarian relief work, or international peace-keeping operations. Although these noncommercial services were given brief mention, they were only summarily described in order that much more focus could be given to the key commercial wireless services that have been described in terms of existing and projected market size as given in Figure 1.5 of Chapter 1.

The Five Drivers model presented in Chapter 1 suggested that telecommunications in general and wireless telecommunications in particular are heavily impacted by the factors of technology, standards, services and applications, management, tariffs, finance, and policy and regulation. It further argues that all of these interact together within the broader marketplace. In short, this model suggests that attempting to understand wireless telecommunications by considering only technology, or only pricing concepts, or only regulation will simply not work. The key to addressing a complex field like telecommunications is thus seen as being at once multidisciplinary and interdisciplinary in scope. It requires understanding of the entire field with a synoptic overview. An integrated set of legal, regulatory, economic, marketing, business, and technological skills is needed to understand this rapidly evolving field.

10.2 THE KEY ELEMENTS OF WIRELESS TELECOMMUNICATIONS

The key elements of each of these areas as addressed in this book are presented in summary form in the subsections that follow.

10.2.1 Market, Services, and Applications

The market that is defined by global telecommunications products and services is expected to reach about $1 trillion (U.S.) by the year 2001 and the wireless component of that market is estimated to be 20 to 25 percent of that total. As can be seen in Figure 1.5 the scope and size of the wireless market will change rather dramatically over the next decade. This is particularly so in terms of shifts in the relative standing of the services provided within this high-growth industry. The sectors projected to grow the most are personal communications service, cellular, specialized mobile radio (including enhanced SMR), wireless LANs and PABXs, and personal digital assistants. Some studies have even suggested that cellular and PCS services alone will capture at least 25 percent of the total telecommunications market in the United States by 2005 and even 20 percent of the global market. These projections, however, seem to be overly optimistic at this time.[1]

Significant growth is also projected for wireless cable television, mobile and direct broadcast satellite systems, as well as fixed and navigational satellite services (including HALE platform technology). Those sectors of the wireless telecommunications market that are projected just to maintain their current status or to lose market share are long-distance microwave relay, short-wave radio telephony, and over-the-air radio and television broadcasting.

In a general sense those wireless technologies that provide mobility, flexibility, and compact transportability will likely thrive and prosper in the marketplace. This is most likely to be so when these wireless systems are also designed to connect to wire-based terrestrial systems in a seamless and highly cost-effective fashion. In short, an information megatrend of the 1990s is that both communications and entertainment will be increasingly mobile and personally interactive.

Wireless services that compete directly with wire-based services for delivery of telecommunications to fixed locations will likely experience little or even negative growth. Those that supplement and augment fiber optic cable systems, on the other hand, can be expected to expand rapidly. This is to say that the so-called Negroponte Flip or the Pelton Merge "effects" will indeed be seen over the next decade and the only issue is the speed and dimension of the transformation. Hybrid systems—those that merge or combine wire and wireless technology—rather than outright substitution (or a dramatic flip-flop) thus seems likely to be the predominant trend for perhaps some decades to come. The next decade will thus be an era of "air **and** glass" as hybrid systems grow and mature.

10.2.2 Technology

The range of technologies that will be developed and deployed in the next decade is long and impressive. The broad level of market support for developments in the wireless field can be expected to promote rapid progress in the following areas.

(a) Advanced digital modulation and encoding will help accelerate advances with regard to many areas such as spread spectrum (CDMA), advanced codec designs for imaging and video, and improved interface protocols and error control techniques. Software advances will, in general, lead hardware advances for some years to come.

(b) Advanced digital compression techniques will likewise generate major gains not only in terms of performance and quality, but also in terms of cost reductions.[2]

(c) Microcellular and picocellular systems within advanced digital mobile networks will create significant increases in frequency reuse and operational capacities. This trend will likely be slower in coming than was first projected as PCS cell sizes are adjusted to accommodate vehicular traffic. SMR digital services will also prove capable of serving over 10 million subscribers.[3]

(d) Advanced node interconnection systems for PCS networks, particularly within cable television and wide band telephone systems, will create new types of networks capable of providing broad band wireless services.[4]

(e) Active reclamation and reallocation of frequencies for new, higher-value, and consumer-defined applications will proliferate. (This may be in terms of reallocations and spectrum auctions or it may be accomplished by indirect means, such as the case of enhanced specialized mobile radio services.)[5]

(f) Advanced satellite systems design concepts will include on-board signal regeneration, signaling, switching, plus cellular beam systems with active phased array antennas.[6]

(g) Advanced intersatellite links with high-speed throughput and multiple satellite interconnection will be achieved with optical link telescopes and millimeter wave systems.[7]

(h) Advanced hand-held and compact ground transceivers will begin to use the latest in strip, patch, and phased array antenna systems. MMIC technologies and digital processing techniques will become operational and the price of these devices should drop dramatically over the next decade.[8]

More fundamental shifts may occur in the longer term. These aspects, as discussed earlier, may include the migration to even higher frequencies, such as the extensive use of infrared transmissions for intraoffice communications or substantial use of millimeter waves for satellite services. The most radical change of all could be in the entire area of frequency use for wireless telecommunications. It has been suggested that the technology now exists for dynamic ranging over many megahertz of frequencies and for choosing temporary frequencies for use on demand. This new technology would allow the practical expansion of total telecommunications capacity over existing static allocation procedures by perhaps two orders of magnitude.

The territorial imperative of existing allocations and licenses for frequency use will not be easily abandoned. This is not only a matter of protecting one's own advantage, but it may also be equally a matter of denying advantage or at least ensuring parity with the competition. A process whereby the entire available spectra, perhaps segmented in two or three parts, is available for dynamic assignment on demand could encourage new applications, lower the cost of telecommunications services, and reduce regulatory processes to a minimum.

One thing seems clear. The world of wireless technology will be driven heavily by the rapid developments in fiber optic cables. The development of soliton pulse technology, the so-called repeater-less cable, and advanced optical switching systems will not only create a surge in the overall field of telecommunications but also challenge the future development of wireless technology as well. We may well see fiber optic cables capable of transmitting a terabit of data per second by 2010. This suggests that communications satellites will likely need to achieve data rates of at least 100 gigabits per second by that time if they are to stay reasonably competitive. Mobile systems to vehicular, maritime, or aeronautical traffic will also likely need to achieve something on the order of 1.5 megabits per second for voice/data/video links to keep pace with projected growth in broad band services.

10.2.3 Economics, Finance, and Administrative Issues

Two factors can contribute most to reducing costs. These are technological innovations and economies of scale or scope. Fortunately for the field of wireless telecommunications both forces are very actively at work. The new technologies identified in the previous subsection coupled with the rapid growth of the subscriber base should certainly continue to push the cost and prices of cellular radio telephones downward. The same is true for GSM services for at least the European and Asia-Pacific markets. Soon PCS services should be following the same trend line of reducing costs and prices. The area of satellite services (including fixed, mobile, broadcast, and navigation) will also show cost and price reductions. These reductions, however, may be more driven by technological innovation than by rapid market growth or major new economies of scale. In fact, it will most likely be a combination of the three.

The other major factor impacting the pricing of wireless telecommunications is that related to competitive forces in the marketplace. In many parts of the world the telecommunications market is being opened to competitive forces. In Europe the licensing process for competitive VSAT terminals is now being opened to competitive suppliers. In many countries such as New Zealand, the United States, and Hong Kong, competitors for cellular or PCS service are being allowed to compete with one another and, in effect, to purchase their frequencies on a competitive basis as well. Some believe that these combined processes including new satellites for personal communications services will produce a mobile service for under $.25 per minute within five years and that ultimately the price differential for both long-distance communications and for mobile services may become very small indeed. This is to say that only value-driven considerations will tend to keep mobile and long-distance service priced significantly above conventional services. In fact, in countries with very low rates for conventional telecommunications such as Mexico and Russia, the cellular rates are already near these low target rates.

This strong pattern of innovation and competition will likely create a dilemma for the field of wireless telecommunications. The potential problem would be that of declining price structure, which without rapid market growth could easily result in lower revenues. This suggests that new services, and especially important new value-added services, will need to be constantly added to the marketplace if net revenues are to continue to increase. Wide band mobile services for video-conferencing and imaging, enhanced GPS controls for "smart highways," mobile telecommunications services for the distribution of electronic newspapers, and "high service/high performance virtual offices" will be among these new value-added services that will fuel new market growth. These high-value and high-productivity services seem to be a predictable part of the future. The highly competitive market for wireless service could also lead to a continuing pattern of merger and acquisition, particularly in the United States. The planned pairing of US WEST, Air Touch, Bell Atlantic, and NYNEX to form a group comparable in size to AT&T's McCaw Cellular is perhaps indicative of this trend.[9]

A final issue is that of financial equilibrium on a global scale. The discrepancy in tariffing of telecommunications services worldwide is enormous. The high rates charged for telecommunications services in countries where there are still monopolies reflect a viewpoint that these markets and services are inelastic. Increasingly, however, call-back systems of other international carriers and competitive satellite systems are serving to provide competition and rates are being driven downward.

The result over the next decade seems likely to be a downward movement of tariffs and greater parity or equilibrium between and among all service providers. In short, satellite, cellular, wire, or other forms of telecommunications will tend to reach common or at least comparatively equal cost and price levels without major discrepancies among them. This will in part be due to the overall driving force of digital processing in all of these transmission technologies. This should also serve to equalize the current discrepancies in the telecommunications rates, which give rise to serious problems and inequalities in the international collections and settlements process. Those entities or countries that seek to hold out against this overall macrolevel trend may very well find themselves losing businesses, investment, and even international standing as a "progressive or enlightened nation."

Innovations in business practices to allow global decentralization of major international corporations as well as broad acceptance of concepts in total quality management and time based management will allow progressive corporations to achieve a strategic advantage based upon consumer satisfaction, rapid innovation and prototyping, and employee responsiveness.

10.2.4 Standards

The currently confused state of international, regional, and national standards-making is not likely to resolve itself quickly. The "balkanization" of the standards-making process if anything continues to accelerate. The combination of business-based, professional society-based, national and regional government-based, and international organization-based

(U.N. and non-U.N.) standards-making organizations have proliferated greatly. Furthermore, the spread of these entities has also led to the spread of special Joint Technical Working Groups and Committees, whose mission it is to try to coordinate the work of their constituent groups.

When there are increasing numbers of joint committees whose responsibilities are defined by coordinating the efforts of standards committees and groups, one senses that the entire process is in trouble. The examples of the PCS standards for the United States and the modulation standards for the so-called "big LEO" system illustrates the serious nature of the problem.

In the first case, the number of options being considered by PCS standards has been painstakingly narrowed from 20 to seven, but then no further resolution among the options proved possible and thus all seven were "accepted as standards." This complicated process has already involved many thousands of weeks of activities and the chances of a uniform global standard for PCS is today almost nil.

In the second case, in deciding on the approved multiplexing technique to be used in the so-called "big LEO" satellite systems the United States—in a process known as negotiated rule-making—took over one year of effort that involved expenses to government and industry in the millions of dollars. Amazingly, this process, despite earnest and well-meant negotiations and technical discovery, ended without a clear-cut result. Some still favored TDMA and others CDMA. The FCC was thus finally forced to segment the band between the two techniques.

The implication of these two examples, plus dozens more that also could be cited, is that both the regulatory and standards-making processes regarding existing and future wireless telecommunications are inefficient, overlapping, and clearly handicapped by too many entities being involved. The processes are also greatly burdened with a high level of overhead, expense, highly legalistic process, and sometimes strong political pressure as well. When it is realized that a measurable percentage of the cost of telecommunications is attributable to standards making and that this cost is now in the billions of dollars, it seems clear that reforms are needed.

The rise of national and regional standards-making bodies as well as numerous international and regional bodies representing professional engineers, industry, and trade-based groups has not only eroded the former strength of the ITU in this area, but also created a process of multiple appeal. If one is not successful in one standards group then one can frequently try again in another forum. If one is a multinational firm operating in many countries the possibilities in terms of promoting one's own approach to a new or revised standard can go on for a very long time. Today there seems to be no easy way to move from regional groups back toward a more integrated international approach or even of reducing the international participants to a smaller and more focused number.

10.2.5 Policy and Regulation

The most significant development in worldwide telecommunications, and most certainly U.S. telecommunications over the last decade, is not a new technology or a service, but rather the creation of a competitive marketplace. The move away from the monopolized provision of telecommunications services, which seriously began in the United States in the 1950s and 1960s, gathered full stride in the 1970s and 1980s . The divestiture of AT&T and the open competition between and among interexchange carriers truly got the competitive process going. This was followed by the rapid introduction of competition within the world of wireless telecommunications on a global scale. This meant the progressive introduction of competitive national, regional, and international satellite systems and the establishment of cellular services on a two per major service area basis. Today there is the clear prospect of a fully competitive future wireless environment within the United States with regard to Enhanced Specialized Mobile Services, AMPS and Digital (IS-55) Cellular Services, and Personal Communications Services. The importance of natural monopolies and economies of scale have proved to be much less important than thought. Today, many regulators are content to say, if a de facto monopoly cannot win out over its new competitors through fair competitive practice then its was not really a "natural" monopoly with true economies of scale after all. Rapid technological innovation makes it truly difficult for any one established entity with a large investment in plant and equipment to dominate any market for any length of time.

The rapid spread of deregulation and competition within the wireless world seems easily explained as simply part of a broader industry trend. The characteristics of wireless telecommunications, however, also makes it particularly well suited to deregulation. This is because new entrants can enter the market without massive capital investment or long-term implementation schedules. The flexibility of coverage of wireless systems allows customers to be easily aggregated from different locations, and capacity can be reallocated to meet emerging market needs. This can, for instance, allow a competitive cellular system to overlay an existing urban terrestrial network as a new competitor. It can also supplement the existing system or act as a hybrid of operationally integrated, fixed, and mobile service provider. In contrast terrestrial networks involve much larger investments and longer-term implementation. The basic need to try to achieve economies of scale, often associated with monopolized terrestrial systems, do not necessarily apply to wireless systems.

The success achieved with competitive wireless systems at the local cellular level, at the regional pager level, as well as at the national and international satellite levels is not only accepted within the U.S. regulatory process, but is also increasingly accepted in other parts of the world. In Europe there are now regulatory policies that encourage competitive licensing and operation of VSAT terminals, open access to public networks for the offering of value-added service, and that create new entities to compete for the provision of public telecommunications services. The force in support of this increased competition is strongest from the European Commission and perhaps less so at the national level. Many new competitive entrants have indeed filed formal complaints with the European Commission to seek to redress alleged difficulties or constraints that they have encountered.

In Japan competition for telecommunications facility provision has been highly structured with only a few Japanese-owned entities allowed to compete in this area. In the area of value-added services, however, open and effective competition by local and international entities has not only been allowed but encouraged. On a worldwide basis there is more and more competition, with over 40 countries moving to implement some level of alternative supply in their telecommunications services.

These changes are also frequently accompanied by streamlined or reduced regulatory control. In particular, entities like OFTEL in the United Kingdom or AUSTEL in Australia have been created to control violations of the framework that allows open and fair competition. This can result in heavy penalties and substantial fines but it also allows a small and streamlined entity to monitor the industry rather than a detailed oversight group to review every service and every tariff, as is the case with the FCC in the United States.

The predominant trend in wireless telecommunications is thus toward more competition, more new participants into the field, and more reliance on the marketplace for regulatory control. Fines for violations of fair competitive practices, controls on predominant carriers, and incentives (such as pioneer preference licenses) represent the type of regulatory environment that is heavily used in the United States today and is likewise being practiced in many parts of the world. This market- and financially driven regulatory environment can have its difficulties in that it is not well equipped to address technical, standards, interoperability, and interconnection issues.

The problems of subscriber privacy and encryption cannot be easily handled by this minimalist approach. Likewise, the issue of the control of childrens' television and the issue of the convergence of the telecommunications, broadcasting, consumer electronics goods, software, and content sectors cannot simply be handled by market forces alone. Social issues that do not easily translate into economic equations cannot be easily handled or even adequately addressed by a regulatory regime whose only responses to marketplace problems are economic equations. Nevertheless, social intervention that runs counter to market demand and economic forces is usually also doomed to failure. Balance between all of the five drivers within the regulatory process is thus often the key to success.

10.3 FINAL CONCLUSIONS

The world of wireless telecommunications is clearly diverse, with dozens of types of operational services now being offered and many more yet to be developed. Despite this diversity the entire field can still be described as having the following characteristics:

(a) frequency demanding and spectrum hungry;

(b) increasingly digitally driven;

(c) significantly constrained by standards;

(d) consumer driven;

(e) demand elastic in terms of price and in terms of new value-added services;

(f) well suited to competitive markets;

(g) chaotically organized with overlapping services, regulatory regimes, standards, and operating systems;

(h) adept at being added as an adjunct to conventional wire-based systems to form potentially seamless hybrid networks;

(i) limited in its ability to compete with high-density fiber optic cable in providing heavy interurban and intraurban services but well suited to complement such services;

(j) driven heavily by the high-end business user (but entertainment- and consumer-defined applications are still important);

(k) well suited to implementation in societies at all levels of economic development and within geographic regions at all levels of population density;

(l) constrained in its deployment by considerations of health-related issues and power-level limits, quality and error control needs, and multipath scatter.

These factors represent an interesting and, at times, even contradictory list of consideration. They could on one hand accelerate the rapid growth and development of wireless telecommunications while inhibiting it on the other. Although not every possible area of wireless telecommunications has been covered in the course of this book, it is hoped that every major aspect of the relevant market, service, technology, finance and management, standards, and policy and regulation has at least been addressed in a useful and meaningful way.

ENDNOTES

(1) Center for Telecommunications Management, "The Telecom Outlook Report" (Chicago, IL: International Engineering Consortium, 1994).

(2) "The Information Wave: Digital Compression is Expanding the Definition of Modern Telecommunications," *Uplink*, Spring 1994, pp. 4-7.

(3) Kozasky, K., "PCS and Redefining SMR," *Wireless*, September/October 1994, Vol. 3, No. 5, pp. 16-17.

(4) "Bell Atlantic Mobile's PCS Now," *Wireless*, July/August 1994, Vol. 3, No. 4, p. 9.

(5) Lockwood, J., "VSAT Network: The New Corporate Solution," *Wireless*, September/October 1994, Vol. 3, No. 5, pp. 28-31.

(6) Starr, L.B., "NASA ACTS: Exploring the Next Generation of Satellite Users," *Wireless*, July/August 1994, Vol. 3, No. 4, pp. 42-43.

(7) NASA/NSF, *Panel Report on Satellite Communications Systems and Technology* (Baltimore, MD: International Technology Research Institute, July 1993), pp. 1-18.

(8) NASA/NSF, *Panel Report on Satellite Communications Systems and Technology* (Baltimore, MD: International Technology Research Institute, July 1993), pp. 20-24.

(9) Zeiger, D., "Four Team Up for Wireless Service," *Denver Post*, October 24, 1994, pp. C-1 and C-10.

PART V

APPENDICES

APPENDIX ONE

SELECTED BIBLIOGRAPHY

The following bibliography is recommended as a basis for further reading and study. As this field is rapidly evolving, it is recommended that current journals in the field be consulted to supplement your readings. This is especially necessary in rapidly evolving areas such as digital wireless technological innovations, new service development, standards, and tariff charges. No text in a field like wireless telecommunications can hope to stay current for any significant period of time.

BOOKS

1993–94 Annual Review of Communications (Chicago, IL: International Engineering Consortium, 1994).

Bowers, R., A. Lee, and C. Hershey (eds.), *Communications for a Mobile Society* (Beverly Hills, CA: Sage Publications, 1978).

Branscomb, A. (ed.), *Toward a Law of Global Communications Networks* (New York: Longman, 1986).

Bruce, R., J. Cunard, and M. Director, *The Telecommunications Mosaic* (London: Butterworth Scientific, 1988).

Calhoun, G., *Digital Mobile Communications* (Norwood, MA: Artech House, 1992).

Cross, T., *Intelligent Buildings* (Boulder, CO: Cross Communications, 1987).

Didsbury, H., (ed.), *The Future: Opportunity, Not Destiny* (Bethesda, MD: World Future Society, 1986).

Hill, A., *Europe's Wireless Revolution* (Chicago, IL: Intertec Publishing, 1991).

Hills, J., *Deregulating Telecommunications* (London: Frances Pinter, 1986).

Howkins, J., and J.N. Pelton, *Satellites International* (London: Macmillan Press, Ltd., 1986).

Inglis, A.J., *Behind the Tube: A History of Broadcasting Technology* (Stoneham, MA: Butterworth, 1992).

Keen, P.G.W., and J.M. Cummins, *Networks in Action: Business Choices and Telecommunications Decisions* (Belmont, CA: Wadsworth, 1994).

Meyers, R. (ed.), *The Encyclopedia of Telecommunications* (San Diego, CA: Academic Press, 1989).

Meagher, C., *Satellite Regulatory Compendium* (Potomac, MD: Phillips, 1993).

Mirobio, M.M., and B. Morgenstern, *The New Communications Technologies* (Boston, MA: Focal Press, 1990).

National Research Council, Computer Science and Telecommunications Board, *Realizing the Information Future: The INTERNET and Beyond* (Washington, DC: National Research Council, 1994).

Newton, H., *Newton's Telecommunications Dictionary* (New York: Telecommunications Library, 1993).

Payne, S. (ed.), *International Satellite Directory* (Sonoma, CA: Design Publishers, 1994).

Pelton, J.N., *The How To Book of Satellite Communications* (Sonoma, CA: Design Publishers, 1992).

——, *Future View: Communications, Technology and Society* (Boulder, CO: Baylin Publications, 1992).

Ramteke, T., *Networks* (Englewood Cliffs, NJ: Prentice Hall, 1994).

Schnaars, S., *Megamistakes* (New York: MacMillan, 1986).

U.S. Congress, Office of Technology Assessment, *Critical Connections: Communications for the Future* (Washington, DC: OTA, 1990).

Williams, F., *The New Telecommunications: Infrastructure for the Information Age* (New York: The Free Press, 1991).

ARTICLES

Abe, G., "The Global Network," *Network Computing*, Nov. 15, 1993, pp. 48-53.

Adams, D., and C. Frank, "WARC Embraces PCN," *IEEE Communications Magazine*, June 1992.

Anderson, H., "The Mobile Professional" (New York: The Yankee Group, 1993).

"AT&T Bell Labs Develops Intelligent Vehicle Highway System Toll-Collection System," *Global Positioning and Navigational News*, Mar. 10, 1994, Vol. 4, No. 3, pp. 1-2.

Aversa, J., "FCCs Okays Satellites Providing Two Way Data Services," *Daily Camera*, Oct. 1994.

Baer, W.S., "New Communications Technologies and Services," in Paula R. Newberg, ed., *New Directions in Telecommunications Policy* (Durham, NC: Duke University Press, 1989), pp. 139-169.

"Band Sharing, 6; Band Splitting 1: Big LEO Report Sent to FCC," *Signals*, Spring/Summer 1993, No. 6, pp. 1-4.

Berger, J., "SMR Shows its Utility," *Wireless*, September 1993, Vol. 2, No. 3, pp. 34-36.

——, "How Do I Get There? Navigation in the 'Smart Car' Age" *Wireless*, May/June 1994, Vol. 3, No. 3, p. 28ff.

Brodsky, I., "Cellular's Multi-Pronged Data Strategy: Making the Right Connections," *Wireless*, July/August 1994, Vol. 3, No. 4, pp. 32-35.

Brown, P., "Business Productivity Through Networking," *INTERNET*, April 1992.

Brown, S., "How the Europeans Respond to Mobile Communications," *Communications News*, May 1994, pp. 28-30.

Bryan, S., "PCN: Prospects in the United States," *Telecommunications*, March 1991, pp. 23-27.

Bushaus, D., "PCS Players Consider GSM," *Communications Week*, October 25,1993.

Carey, J., and M. Moss, "The Diffusion of New Telecommunications Technologies," *Telecommunications Policy*, June 1985, pp. 145-158.

Carnevale, M.L., "Broadcasters Gain Support for Measure to Open Spectrum for New Services," *Wall Street Journal*, March 1, 1994, p. B6.

Carraway, R.L. , J.M. Cummins, and J.R. Freeland, "The Relative Efficiency of Satellites and Fiber-Optic Cables in Multipoint Networks," *Journal of Space Communications* (Amsterdam, Netherlands: IOS Press, January 1989, Vol. 6, No. 4, pp. 277-289.

Cauley, L., "MCI's Entry Adds New Dimension to Wireless Race" *Wall Street Journal*, March 1, 1994, p. B4.

"CDPD—Here, There and Everywhere?" *Wireless*, June 1993, Vol. 2, No. 2, p. 6.

Channing, I., "Customers Wanted: Prospects for Personal Communications Services," *Communications International*, September 1993.

Channing, I., and R. Burr, "Worldwide Digital Cellular," *Mobile Communications International*, Winter, 1994.

Tsoi, K.-C.A., "User Interface Issues for Cellular Phones," *1993–94 Annual Review of Communications* (Chicago, IL: International Engineering Consortium, 1994), pp. 679-681.

Cole, L.S., "Phoenix Police Take Command of Wireless Technology," *Wireless*, September 1993, Vol. 2, No. 3, pp. 52-54.

"The Communicopia Study: C-5 Convergence" (New York: Goldman-Sachs, 1992).

"Comsat Appeals for Privatization," *Satellite Communications*, April 1994, pp. 10-11.

Costin, G., "Cellular in the Two-Way Marketplace: Bonanza or Bust," *Mobile Communications Business*, September 1986.

Davis, A., "Cable Overbuild: Alternative Video Access Opportunity," *1993–94 Annual Review of Communications* (Chicago, IL: International Engineering Consortium, 1994), pp. 115-118.

Davis, H., "Enhanced SMR: The First PCS?" Telestrategies PCS Conference, Washington, DC, June 1994.

"The DBS Market," *Via Satellite* (Potomac, MD: Phillips Publishing, 1993).

Deloitte and Touche, "1993 Wireless Communications Industry Survey" (Atlanta, GA: Deloitte and Touche, 1993).

"Desperately Seeking PCS Standards," *Data Communications*, December 1993.

"Electronic Privacy Bill Passes House Committee," *Telephony*, May 26, 1986.

DeSilva, E.W., "PCS/Wireless Regulatory Update," *ICA Expo*, Dallas, TX, May 1994.

"The Direct TV DBS System," *Via Satellite* (Potomac, MD: Phillips Publishing, 1993).

Ducey, R.V., and M.R. Fratrik, "Broadcasting Response to New Technologies," *Journal of Media Economics*, Fall 1989, p. 80ff.

Dyer, J.E., "Disaster Recovery in a Cellular Service," *1993–94 Annual Review of Communications* (Chicago, IL: International Engineering Consortium, 1994), pp. 613-614.

Frieden, R., "Wireline vs. Wireless: Can Network Parity Be Reached?" *Satellite Communications*, July 1994, pp. 20-23.

Glaser, P., "The Practical Uses of High Altitude Long Endurance Platforms" (Graz, Austria: International Astronautical Federation), October 1993.

"Global 2000 Report on Telecommunications" (Washington, DC: National Telecommunications and Information Administration, 1990).

"The Green Paper on Satellite Communications, European Commission" (Brussels, Belgium: The European Commission, 1992).

Grant, M., "UPT: Can the Phone Take the Strain," *Communications International*, August 1993.

Gustafsson, A., "Lessons Learned from Launching a GSM Digital Cellular System in Sweden," *1993–94 Annual Review of Communications* (Chicago, IL: International Engineering Consortium, 1994), pp. 618-621.

Hardy, T., "Personal Communications Services," *IEEE Communications Magazine,* June 1992.

Horwitz, C., "The Rise of Global VSAT Networks," *Satellite Communications* (Denver, CO: Argus Publishing, April 1994), pp. 31-34.

"The INTELSAT Agreement and Operating Agreement," TIAS Series (Washington, DC, U.S. State Department, 1992).

"ITU International Table of Frequency Allocations," as contained in Part 2 of the FCC Rules and Regulations (Washington, DC: Government Printing Office, 1994).

"The Japanese Automatic Vehicle Location Market Continues to Gain Momentum," *Global Positioning and Navigational News,* March 10, 1994, Vol. 4, No. 3, reference files.

Kachmar, M., "The Goal Is Control of Time and Place," *Wireless,* November 1993, Vol. 2, No. 4, pp. 22-23.

Ketchum, J., "Cellular Mobile Telecommunications," Tutorial on Wireless Technologies, Supercom Conference, Houston, TX, 1991.

Kirvan, P., "Implementing a Wireless PBX," *Wireless,* June 1993, Vol. 2, No. 2, pp. 34-38.

"Ku-Band Sharing," *Satellite Communications,* April 1994, p. 12.

Lang, R., and J. Sauer, "Scalable Dense Wave Division Multiplex Photonics for an All Optical Network," White Paper, Spectra Diode Laboratory, San Jose, CA, 1994.

Lindstrom, A., "Nextel Introduces First U.S. Digital Cellular Network Based on GSM," *Communications Week,* October 4, 1993.

Lipman, A., "Leaping from the Barricades," *IEEE Communications Magazine,* June 1992.

Lucas, J., "PCS vs. Cellular," *Telestrategies* (McLean, VA: Telestrategies, August 1993), pp. 1-12.

Malone, R., "Wireless Communications With a Human Face," *Wireless,* September 1993, Vol. 3, No. 4, pp. 12-14.

Manuta, L., "Big Leo Equals a Big Deal," *Satellite Communications,* April 1994, pp.14-15.

Markus, M.L., "Toward a Critical Mass Theory of Interactive Media: Universal Access, Interdependence and Diffusion," *Communications Research,* October 1987, pp. 491-210.

Marshall, P., "Global Television by Satellite," *The Journal of Space Communications and Broadcasting,* January 1989, Vol. 6, No. 4.

Mason, C., "Will the U.S. Remain Competitive in the Wireless Future?" *Telephony,* July 12, 1993.

McCaw, C.O., "Cellular Communications," *1993–94 Annual Review of Communications* (Chicago, IL: International Engineering Consortium, 1994), pp. 43-44.

———, "Changing the World," *Wireless,* June 1993, Vol. 2, No. 2, pp. 23-26.

"MCI Is Filling Up Its Dance Card," *Business Week,* October 24, 1994, pp. 60-61.

"Motorola's Integrated Phone-Pager," *Wireless,* September 1993, Vol. 2, No. 3, p. 55.

"Motorola's Pioneer Preference Applicaton for Iridium Loses," *FCC Week,* September 1992.

Mouly, M., and M.B. Pautet, "The GSM System for Mobile Communications" (Palaiseau, France: Bruneau, 1992).

Nakonecznyj, I.T., "The Wireless Revolution—It's Here Almost!" The NEC ComForum, Orlando, FL, 1992.

Omura, J., "Improving Network Security: User Authentification and Encryption," *Telecommunications,* May 1991.

Ormerod, J., and A. Butterworth, "The Ideal Communications Service," *IEEE Communications Magazine,* June 1992.

Palumbo, W.J., "Wireless PBX Access," *ICA Expo,* Dallas, TX, May 1994.

Paschall, L.M., "Security Aspects of Satellite and Cable Systems," *Journal of Space Communications* (Amsterdam, Netherlands: IOS Press, January 1989), Vol. 6, No. 4, pp. 269-276.

Pelton, J.N., "Five Ways Nicholas Negroponte is Wrong About the Future of Telecommunications," *Telecommunications,* April 1993, Vol. 11, No. 4.

——, "The Globalization of Universal Telecommunications Services," *Universal Telephone Service: Ready for the 21st Century?* (Wye, MD: Institute for Information Studies); *Annual Review,* Aspen Institute, November 1991, pp. 141-151.

——, "How INTELSAT Was Privatized While No One Was Looking," *Via Satellite,* February 1989.

——, "Toward a New National Vision for the Information Highway," *Telecommunications,* September 1993, Vol. 11, No. 9.

——, "Will the Small Satellite Market Be Large?" *Via Satellite,* April 1993.

"Public's Privacy Concerns Still Rising," *Privacy and American Business* (Hackensack, NJ: Center for Social and Legal Research, September 1993).

"Reformed ITU Filing Procedures—Brokers for Orbital Space-Boon or Bane?" *Via Satellite,* May 1994.

Ross, I., "Uniform Standards Are Required to Ensure Global Cellular Growth," *Mobile Radio Technology,* May 1986.

Schnee, V., "Giant EDS Spreads Wireless," *Wireless,* September/October 1994, Vol. 3, No. 5, pp. 18-22.

Solomon, R.J., "Shifting the Locus of Control," *Annual Review of Communications and Society* (Queenstown, MD: Institute for Information Studies, 1989).

Special Edition on Mobile and Small Satellites: *Journal of Space Communications,* April 1993, Vol. 10, No. 2.

Starr, L.B., "NASA ACTS: Exploring the Next Generation of Satellite Users," *Wireless,* July/August 1994, Vol. 3, No. 4, pp. 42-43.

Strege, A.F., "Developing Global Wireless Systems for PCS," *AT&T Technology,* Autumn 1993.

"Studies Support Speedy Licensing," *Inside Wireless,* Vol. 2, Issue 7 (Englewood, CO: Four Pines Publishing, April 27, 1994), pp. 3-4.

Tarlin, J., "Cellular Penetrates the World," *Pan-European Mobile Communications,* Autumn 1992.

——, "Intelligent Base Stations for GSM," *Pan-European Mobile Communications,* Autumn 1992.

Taylor, J., "Trial by Auction: The Greening of PCS," *Wireless,* 1993, Vol. 2, No. 2, pp. 32-34.

——, "PCS in the U.S. and Europe," *IEEE Communications Magazine.*

Taylor, L., "PCS Frequency Auction," *Signals,* November 1993.

"The Information Wave: Digital Compression is Expanding the Definition of Modern Telecommunications," *Uplink* (Los Angeles, CA: Hughes Aircraft, Spring 1994), pp. 4-7.

Titch, S., "Cellular Plight Will Last Until the 1990's," *Communications Week,* February 2, 1987.

Toll, D., "The Promise of CDPD," *Wireless,* March 1994, Vol. 3, No. 2, pp. 28-33.

Vorick, F.L., "Cellular Service Evolution," *1993–94 Annual Review of Communications* (Chicago, IL: International Engineering Consortium, 1994), pp. 688-691.

Wainwright, R.A., "Quality as a Competitive Edge," *1993–94 Annual Review of Communications* (Chicago, IL: International Engineering Consortium, 1994), pp. 939-943.

Weber, J., "Motorola's Rapidly Developing Peer-to-Peer Network is Helping to Achieve A World-Wide 'Wall-Less Workplace,'" *Networking Management,* July 1992.

Whitehead, J., "Cellular System Design: An Emerging Engineering Discipline," *IEEE Communications Magazine*, February 1986.

Williams, D., "Looking to the Future: The Strategic Development of GSM," *Mobile Communications International*, Winter 1994.

Wimmer, K., and B. Jones, "Global Development of PCS," *IEEE Communications Magazine*.

"World's Top 10 Markets for Telecommunications Investment Forecasted," *Global Telecom Report*, April 4, 1994, Vol. 4, No. 7, pp. 1-2.

Zehle, S., "The Key to EC Cellular Communications," *Communications International*, January 1990.

Zeiger, D., "Four Team Up For Wireless Service," *Denver Post*, October 21, 1994, pp. 1C and 10C.

REPORTS AND DOCUMENTS

Center for Telecommunications Management, *The Telcom Outlook Report* (Chicago, IL: International Engineering Consortium, 1994).

Electronics Industry Association, "EIA/TIA Interim Standard for Cellular Radio-Telecommunications Intersystem Operations: Functional Overview," Interim Standard 41.1-B,

European Telecommunications Standards Institute, GSM Recommendation 2.07, "Specification of the Mobile Station Features," March 1990.

European Telecommunications Standards Institute, GSM Recommendation 2.09, "Specifications of Security Aspects."

Federal Communications Commission, "Regulatory Treatment of Mobile Services: Report and Order on Personal Communications Services" (Washington, DC: Adopted June, 1994).

Federal Communications Commission, "Regulatory Treatment of Mobile Services: Third Report and Order Related to Specialized Mobile Radio Services" (Washington, DC: Adopted August 9, 1994 and released September 23, 1994).

International Telecommunication Union, "ITU-T Recommendation F.850—Principles of Universal Personal Communications, March 1993.

International Trade Commission, "Global Competitiveness of U.S. Advanced Technology Industries: Cellular Communications," Publication 2646, Washington, DC, June 1993.

NASA/NSF, *Panel Report on Satellite Communications Systems andTechnology* (Baltimore, MD: International Technology Research Institute, July 1993).

APPENDIX TWO

GLOSSARY OF TERMS

A/B Switch Permits   This is a feature that applies to roaming when the subscriber is outside the normal service area. It allows selection of either the wireless system (System A) or the wire–line system (System B).

Active Phased Array Antenna   An antenna with integrated electronic components that are programmed to operate in phase with one another to increase the overall performance or gain of the antenna. These antennas also are able to track a mobile transmitter, including a satellite, to maintain signal by "pointing electronically" to the moving source.

American National Standards Institute (ANSI)   This is the major standards-making institution of the United States. Within this group the T-1 Committee has primary responsibility for telecommunications related standards.

Advanced Mobile Phone Service (AMPS)   This is the original standard for analog cellular mobile telecommunications.

Analog Transmission   A method of telecommunications that operates on the basis of a "representation" or analog of the sound or image that is to be transmitted.

Asynchronous Transfer Mode (ATM)   A fast packet-switching protocol based upon the relays of fixed-sized cells. This is the basic technology for broad band ISDN switching and signalling.

Band-width   The amount of spectra available for a particular telecommunications function. It is literally computed by subtracting the lowest frequency in a band from the highest frequency in the band.

Base Station   In a cellular mobile telecommunications system there is typically a station at or near the center of a cell that relays all signals to and from the mobile subscribers. This unit, called a base station, is connected by wire, coaxial cable, fiber, or microwave relay to a Mobile Telecommunication Switching Office, which services several cells.

Base Station Controller This controller maintains the overall direction and operation of a group of Base Transceiver Stations (BTS) that are located within the cell sites.

Base Station Transceivers This is the unit that is responsible for the transmitting and receiving of the radio frequency signals between the cell-site controller and the mobile units in the cell's area of coverage.

Bellcore The collective research and requirements-setting group which supports all of the seven Bell operating companies. This was created as part of the AT&T divestiture process by splitting up Bell Labs in order to create this new unit.

Broadcast Satellite Service (BSS) In the official naming system of the International Telecommunication Union, direct broadcast satellite systems that provide direct service to the home are designated as the Broadcast Satellite Service (BSS). This is normally for television services, but some BSS satellites lease or provide channels for direct broadcast radio as well. These may be in the regular BSS bands or a special BSS Radio band.

Broadcast Satellite Service Radio (BSSR) There is a separate allocation for direct broadcast radio service that is exclusively to broadcast radio directly to the home or to vehicles. For the United States this is likely to be in the 2.2-GHz band.

Bus A "bus" in the field of satellite communications is the basic support structure and power system that provides a platform for the satellite telecommunications subsystem. In terms of wireless communications within the office it is the unit that provides total coverage of an office unit by RF signals and maintains switched interconnection with the overall telecommunications network. This is sometimes called a data bus, although in the future it will deliver all forms of telecommunications.

Call In Absence Alert This is a system that displays all attempted in-coming calls when the subscriber was out of service range, was away from the transceiver phone, and so on.

C Band This is the frequency band in the range of 4–7 GHz. It is among the primary bands used for satellite communications with 3.7–4.2 GHz being used as the down-link band and 5.7–6.2 GHz being used as the up-link band. This band is now heavily saturated by existing users of satellite communications. Most new satellite systems now use the higher K band frequencies.

CCIR The Consultative Committee on International Radio of the International Telecommunication Union. Shortly the radio and terrestrial telecommunications functions will be integrated in the new ITU structure.

CCITT The Consultative Committee on International Telephone and Telegraph of the International Telecommunication Union. Shortly the radio and terrestrial telecommunications functions will be integrated in the new ITU structure.

Cell The geographically restricted area, often hexagonally shaped units, into which a mobile telephone system is divided. Through control of the power levels and pointing of the antennas at the center of the cell, low-power radio signals cover the entire cell, but with a minimum of overlap into the adjacent cells.

Cell-Site Control The computer-controlled program that manages the assignment and hand off of channels within a cell.

Cellular Digital Packet Data (CDPD) A new system implemented in 1994 for sending data efficiently over an analog cellular system.

Cellular PC A wireless transmit-and-receive capability built into a personal computer in the form of modem that allows communications, usually in an available unlicensed band.

Cellular Radio Telephone Service The basic cellular radio service used in the United States today is in the 900-MHz band and uses analog modulation for transmitting and receiving signals within cell sites typically some one to two miles in radius.

Circuit Switching A switching method based upon dedicating a channel to a voice conversation for the duration of the call. See packet switching.

Closed Architecture A closed or proprietary system that is not open to interconnection with other networks.

Code Division Multiple Access (CDMA) This advanced modulation technique, also known as "spread spectrum" reuses the spectrum many times over by overlaying signals in a broad frequency band and then using unique digital codes to extract the wanted signal from the other unwanted signals.

Codec The term codec is a combination of the words coder and decoder. These are used for digital encoding and decoding of analog signals. They are required to derive digitally compressed signals such as 56, 128, 256, 384, or 512 kilobit per second video signals.

Data Compression This is a process of trying to send information more efficiently in the digital format but reducing "unneeded" information. This is usually done by reducing either the bandwidth, the time needed for generation or storage of information, or redundancies in the information. Techniques that are employed include bit mapping, null suppression, pattern substitution, or code-book pattern matching.

DBS Direct Broadcast Satellite. These high power satellites can broadcast directly to the home or office. Examples of DBS satellites include the DirecTV system of Hughes, the N-Star system of Japan, TVSAT of Germany, TDF Satellite of France, and British Sky Satellite of Europe.

DBSR Direct Broadcast Satellite Radio. This is a different type of satellite designed to broadcast at lower-band radio frequencies directly to homes, offices, and vehicles. This will be a CD-quality service that will also provide security and other service features. Only a limited number of satellites have been licensed for this service and the market is not yet proven.

Decibel (dB) A way of measuring power levels in telecommunications systems based upon a logarithm scale. Each increase of power by 3 dB represents a doubling of power and each decrease of power by 3 dB represents a halving of power.

Decoder A device that converts a digitally encoded signal to analog form. It is often paired with an encoder to create an encoder/decoder or codec.

dBW Power levels measured in decibels with reference to one watt.

Digital Cellular This is the latest form of cellular service. Rather than using FDMA as is used with conventional cellular, digital cellular will typically use either TDMA or CDMA modulation. This much more efficient form of service will be able to derive from 5 to 15 times more capacity from the equivalent frequency band.

Digital Echo Canceller A digital echo canceller in a voice circuit detects incoming echo and other spurious signals and generates an offsetting signal to exactly cancel out the unwanted wave forms. These devices are much superior in performance to earlier devices known as echo suppressors.

Digital Signal Processing Essentially this is to adapt general-purpose microprocessors for specialized telecommunications purposes such as echo cancellation, call progress monitoring, voice processing, digital compression, and so on.

Down-link The portion of a complete satellite link that connects the satellite to the ground segment antenna.

Echo Canceller See Digital Echo Canceller

e.i.r.p. Effective Isotropically Radiated Power. A systematic and standardized measure of power in telecommunications systems.

Electronically Steerable Phase Array Antenna This is essentially the same as an Active Phase Array Antenna as defined previously.

Electronics Industry Association This is a trade group of the electronics manufacturing industries in the United States who promote trade in electronics products and develop industry standards. The have played a key role in the development of the new Consumer Electronic Bus (CE Bus) standards, for instance.

Encoder See Codec or Decoder.

Encrypting The cyphering of a message, video image, voice signal, or some other form of communications by applying a mathematical algorithm to a digital signal for purposes of security. This is a much more sophisticated and difficult-to-decode process than simple scrambling of signals, as is often done in cable television systems.

Error Control A process of checking that a message has been successfully transmitted to the intended location and to have any incorrectly transmitted information resent. This is particularly crucial for transmission systems with high noise levels or bit-error rates less than one in one thousand.

European Telecommunications Standards Institute (ETSI) This is an institution created within the framework of the European Commission to coordinate the adoption and implementation of telecommunications standards uniformly throughout Europe.

Extremely High Frequency (EHF) This is the frequency range from 30 to 300 GHz, which is sometimes also referred to as the millimeter wave band. These frequencies with extremely small wavelengths (e.g., millimeter) are not heavily utilized because of the very high level of rain attenuation that occurs in this range. Research activities in the 30–60 GHz band are being undertaken in Japan and the United States.

Fairness Doctrine This is a rule enforced by the FCC that stipulates that both sides of an economic, social, religious, or political issue must be addressed by broadcasting organizations. The print media is, however, not held subject to the same strict standards by somehow being held to have greater latitude in terms of the application of the First Amendment.

Federal Communications Commission An independent regulatory agency for the field of communications whose existence is stipulated under the Communications Act of 1934 and its amendments. This commission is to have balanced partisan representation and is to regulate broadcasting, cable television, telecommunications carriers, and equipment suppliers by overseeing tariffs, competitive structures of the industry, licensing of frequencies, and industry standards. The FCC Commissioners are appointed by the President and confirmed by Congress. Currently there are five FCC Commissioners.

Fixed Satellite Service (FSS) This is the predominant satellite market today. This service connects fixed earth stations together. This is typically accomplished via a satellite in geosynchronous earth orbit and primarily in the C and Ku Band frequencies. There are over 200 geosynchronous satellite systems provided FSS services on a global, regional, or national scale. Most national FSS service consists of video distribution or very-small-aperture terminal (VSAT) corporate networks.

Flat Antenna For many years antenna shapes have tended to be paraboloid in design so as to provide the maximum catchment of incoming radio signals and to provide the maximum focus for transmitted signals. With the advent of phase array antenna systems the capture of incoming signals and the forming of outgoing signals can be "electronically shaped" rather than physical shaping, as in conventional antennas. Thus flat antennas composed of phase array elements can be programmed to achieve such tasks as tracking of moving satellites, create multiple transmission beams, and so on.

Frequency   The number of oscillations per second that a modulated radio wave makes. The shorter the wavelength of a radio signal the higher its frequency.

Forward Error Correction   One of the most common forms of error control methods used in digital transmission systems is that of forward error correction. In this system the accuracy and destination of a message are checked along the transmission path and retransmissions are provided if errors are detected. This system is frequently used in satellite system transmissions. Today with ATM switching system in very-low-bit-error rate fiber optic systems the reliance on FEC is being dramatically decreased with often only one end-to-end check on transmission accuracy.

Frequency Division Multiple Access (FDMA)   This is the most common analog channel-sharing scheme that is used in conjunction with FM modulation. The information is modulated using the frequency characteristics to contain the key information. This is more tolerant of noise and interference than amplitude modulation.

Future Public Land Mobile Service (FPLMS)   This is an ITU-sponsored standard for proposed worldwide digital mobile. Ultimately it is hoped that mobile and fixed terrestrial telecommunications services could be integrated into a totally integrated service known within the ITU as the Universal Personal Telecommunications Service (UPT).

Geosynchronous Earth Orbit (GEO)   This is the orbit that is some 35,870 kilometers or 22,230 miles above the earth's surface. It is at this exact orbital location that a state of dynamic equilibrium is achieved between the earth's gravitational pull, the orbital momentum of the satellite, and the velocity that allows the satellite to make exactly one revolution every 23 hours and 56 minutes. This because of celestial mechanics is exactly equal to the earth's rotation below. Thus a satellite in this orbit, which is almost one-tenth of the distance to the moon, appears to be synchronous to the earth's surface. If the satellite remains exactly in the equatorial plane, then the satellite can also be said to be geostationary as well.

Geosynchronous Satellite System (GSS)   Satellites that operate in the geosynchronous orbit are referred to as Geosynchronous Satellite Systems. Most of these are telecommunications satellites but they can also provide surveillance, meteorological, navigational, and remote-sensing services as well.

Ghosting   The transmission of radio waves in free space can result in reflections of signals off trees, vegetation, buildings, mountains, or other objects. This can create multiple signals being received by an antenna. In the case of television signals this can create a visual effect sometimes called ghosting. The same phenomena, known technically as multipath, can also serve to degrade the performance of mobile radio services.

Global System for Mobile Communications (GSM)   Originally this French-developed digital standard for mobile communications was called Groupe Speciale Mobile. This standard, which is based on TDMA, is now the Pan-European digital cellular standard and is being widely adopted in the Asia-Pacific region as well. Telepoint 2 is considered a derivative of GSM.

Hand-off   Within the cells of a mobile telephone system sensors monitor the movement of subscribers employing a system that detects increasing or decreasing power levels. This process anticipates when someone's vehicle will leave one cell and enter another so as to hand off the signal from one cell to another at the optimum time.

High Frequency   This is the band in the frequency band from 3 to 30 MHz. It is also referred to as shortwave. This band is used for international radio broadcasting, amateur radio, and many other purposes.

High-Tier Mobile Service This is a Personal Communications Service that can provide a wide range of services and, in particular, provide both pedestrian and vehicular service. Low-tier mobile service is typically a pedestrian-based service only.

Home Location Register Database wherein the local mobile subscribers' addresses and other key information is stored.

Improved Mobile Telephone Service (IMTS) A standard of mobile telecommunications implemented prior to cellular telephone service. It allows private channel assignments and direct dial service without using a dispatching operator.

Intermediate Frequency (IF) The medium-wave radio frequencies that are used in satellite systems as well as in LANs as the translating frequencies to place signals within RF carrier waves.

International Mobile Subscriber Identity (IMSI) This is an individual assigned code for GSM service that enables the system to identify key information about the user, such as the subscriber's country, the Public Land Mobile Network designation, Home Location Register, and so on.

Intersatellite Link (ISL) There are a number of applications wherein direct communications between two or more satellites is desirable. These can be accomplished at different frequencies and different types of ISLs can be much more demanding than other types, particularly if the link is short in range or very long range. The three principal types of ISLs are: (a) low earth orbit to geosynchronous, usually for the purpose of data relay; (b) short-range ISLs between reasonably contiguous satellites either in GEO, MEO, or LEO orbits; and (c) long-distance interregional satellite links over spans measured in thousands of kilometers. The ISL links can be established via radio frequencies in the SHF or EHF range or by optical links. In some low earth orbit systems that have been proposed, a satellite could be connected to up to eight other satellites by ISLs at any one time.

IS-41 The EIA and CTIA standards for interconnect of wireless mobile communications systems.

IS-54 The EIA and CTIA standard for analog communications in the United States.

IS-55 The EIA and CTIA standard for TDMA-based digital telephone service in the United States, which is based upon an upgraded version of the AMPS analog standard.

ISDN This stands for Integrated Services Digital Network. It represents a unified set of standards that are designed to ensure high-quality standards for digital transmission. ISDN covers all types of applications that can be sent through digital channels on an integrated basis. The basic transmission unit is a B channel or bearer channel that operates at 64 kilobits per second or multiples thereof. The signaling for ISDN is provided through a separate data channel known as a D channel.

ISDN-Basic Rate The lowest common denominator for ISDN is a Basic Rate Interface consisting of 2B + D channels. This is equivalent to two 64 kilobit bearer channels and one 16 kilobit signaling channel or 144 kilobits per second.

ISDN-Primary Rate The Primary Rate Interface operates in the United States at the rate of 23 B + enhanced D Channel. This is equivalent to 1.5 megabits per second. In Europe it is 31B + D or 2 megabits per second. The frequency restrictions that apply to mobile cellular service make it difficult for full ISDN standards to be achieved, but switching interfaces with the PSTN network have been successfully achieved in most instances.

JCD Japanese standard for mobile telephone services. This is now giving way to the more pervasive GSM global standard.

Joint Experts Meeting The joint effort of the American National Standards Institute (ANSI) and the Telecommunications Industry Association (TIA) to develop U.S. standards for the new PCS service.

K Band The broad frequency range from 10.9 to 36 GHz that covers the high microwave/low millimeter range. This range covers both the Ka and Ku bands used for satellite communications.

Ka Band The upper microwave band from 20 to 30 GHz that is increasingly being used for satellite communications. Experimental satellites that have been recently deployed to develop this new band include the U.S. NASA project known as ACTS, the Japanese Experimental Test Satellite VI, and the European Space Agency Artemis project.

Key for Authentication The subscriber authentication key is known as (Ki). It is a fixed, secret, and personal key that is used to authenticate the subscriber upon accessing the network.

Key for Ciphering This ciphering key is known a Kc. It is used to cipher and decipher the data transmitted over the radio channel and thus provides confidentiality for the free-space transmission. The Kc remains valid for only one data session.

Ku Band The microwave band in the 12–18 GHz range that is used for satellite communications and DBS services.

L Band The band that includes the 1600 MHz/1500 MHz frequencies that are used for mobile satellite communications.

Local Loop A local connection that can be wire or wireless that connects a home or office to a larger telecommunications network.

Location Area Identification This is the database that is accessed to identify where a mobile subscriber is currently located. This information is maintained in the mobile station as well as Location Registers (i.e., home and visiting).

Look Angle The angle at which a ground transceiver "looks" to a satellite to which it is operational. For FSS service this is usually no lower than 5 degrees and for MSS service this is usually no lower than 30 degrees.

Low Earth Orbit (LEO) Satellite deployed in close proximity to the earth's surface often between 700 and 2000 kilometers altitude. These can be "little LEOs," which operate at frequencies typically below 1 GHz and that provide store and forward services in the global constellations of about 12 satellites per system. Satellites in low earth orbit can also be "big LEOs," which typically operate between 1600 and 2200 GHz. These can provide voice and data services but require some 40 or more satellites to provide continuous coverage of the entire world. Recently a mega-LEO satellite with some 840 satellites in the global constellation has proposed the use of Ka band frequencies to provide broad band services.

Low Frequency The radio frequencies between 30 and 300 KHz. The very limited band-width in the LF band limits the practical use of this frequency for any wider band purpose. Navigational and naval communications applications are located in this band, among others.

Masking Angle The angle of satellite transmission to the earth's surface. It is an inverse concept to an earth station's look angle to the satellite.

Medium Earth Orbit (MEO) Medium earth orbit satellites are typically used for remote sensing or mobile communications. A typical orbit is highly inclined and has about a 8000–12,000 kilometers apogee (or high point above the earth). Rather simplistically they are higher than low earth orbit systems and lower than geosynchronous satellite systems.

Microcellular Communications Service Conventional cellular radios have generally used cell sizes measured in several square miles or kilometers and sufficient power to communicate with vehicular traffic. The new digital PCS service will operate with lower power and much smaller cell coverage zones. These may be only a fraction of conventional cellular. These cells measured in only a fraction of a square kilometer or mile have been called microcells and the service has sometimes been described as microcellular communications service.

Microwave or Millimeter Wave Bus An RF-based communications module for intraoffice communications that uses microwave or millimeter frequencies for this purpose. Today most of these services are offered in the UHF band at below 2 GHz in the Part 15 unlicensed frequencies and the soon to be allocated unlicensed PCS frequencies. Also see bus.

Microwave Frequency The microwave frequency band is generally equivalent to the Super High Frequency Band of 3–30 GHz.

Microwave Relay A terrestrial relay usually mounted on a building, a tower, or a mountaintop that is used in the transport of telecommunications signals over long-distance routes. Due to the curvature of the earth these relays typically are spaced some 30–40 miles (or 50–65 kilometers) apart. In developed countries, microwave relay is being replaced by fiber optic cable because of increased capacity, higher-quality performance, and lower operating costs.

Millimeter Wave Frequency This is also known as the Extremely High Frequency (EHF) band. It ranges from 30 to 300 GHz. It is the highest radio frequency currently in commercial telecommunications use.

Mobile Telecommunications Switching Office (MTSO) This is the key switching and signaling facility that services several geographically proximate cellular telecommunications cells. (See illustration in Chapter 4.)

Modulation The process of altering a radio or optical signal so as to send information over some telecommunications transmission system. The primary analog modulation systems are amplitude modulation, frequency modulation, and single side band. The primary digital modulation system are time division multiple access and code division multiple access.

Modem This is a combination word that stands for modulator/demodulator. This device modulates and demodulates telecommunications signals.

Multipath When a wireless signal is irradiated into space it is reflected and refracted as it encounters obstacles such as mountains, buildings, trees, and vegetation. These signals are thus diverted and bounce back or forward so as to reach a distant receiver after the direct signal. This creates multiple pathways to the receiver and interference to the primary signal. This is called multipath interference or ghosting in television signals.

Narrow Band PCS This is the narrow band paging and messaging service that was allocated by the FCC in 1994. The frequencies that were auctioned off at a cost of about $500 million include: 900–901 MHz, 930–931 MHz, and 940–941 MHz.

National Telecommunications And Information Administration (NTIA) This is the agency within the U.S. Department of Commerce that manages the government's frequency allocations, conducts research on telecommunications standards, and assists in defining the U.S. strategic planning process for telecommunications. Its primary offices are in Washington, D.C. and Boulder, Colorado.

National Institute of Standards and Technology This is the former U.S. Bureau of Standards. It promotes the development of new technology through its Advanced Technology Program (ATP) and it maintains the U.S. technical standards. Its primary offices are in Gaithersburg, Maryland and Boulder, Colorado.

NMT-450 and NMT-900 Swedish standard for mobile telephone service that is now giving way to the more pervasive GSM standard.

Omniantenna This is an nondirectional antenna, often a dipole, which is capable of collecting signals from all directions. Because the antenna is nonfocused, it is low in gain and sensitivity but it has the advantage of detecting a signal from any incoming direction. These are frequently used in mobile communications systems.

On-Board Processing   This is a satellite that has the ability to process an incoming digital signal and completely regenerate it for retransmission back to earth. This is in contrast to a conventional "bent-pipe" satellite, which can only filter and amplify the incoming signal for retransmission.

Open System Interconnection (OSI)   This is the set of standards developed by the International Standards Organization (ISO) and the CCITT of the ITU to allow all telecommunications systems to interconnect using the globally approved seven-layer protocol used with the CCIS Number 7 Signaling System.

Packet Switching   A switching method whereby information is divided up into small packets and then flexibly routed over a network on a flexible basis. The message is then reassembled at the receiving location.

Pager   A one-way remote receiver system that can notify a subscriber that they should call their home base or can display simple digital text or voice-mail messages.

Pager Display   A digital display that allows a pager to indicate short multibyte messages. Typically this is a liquid crystal display.

Parabolic Antenna   A directional high-gain antenna with a parabolically shaped reflector designed to capture the maximum amount of radio waves from a distant transmitting source. These types of antennas are frequently used in satellite communications.

Personal Communicator   A multipurpose remote, mobile communications device capable of being carried by a person. Its minimum functionality includes integrated voice communications, paging, two-way electronic messaging, and facsimile. It is typically capable of being "awakened" from a sleeping low-power mode to receive messages.

Personal Communications Network (PCN)   This is a network that offers Personal Communications Services. See PCS below.

Personal Communications Services (PCS)   A new form of digital communications service operating near and below the 2-GHz band. The broad band PCS allocation includes a 120-MHz reallocation of frequencies between 1850 and 1990 MHz. PCS involves smaller and less powerful cells than conventional cellular service. There is both a narrow band and a broad band allocation of PCS plus a narrow unlicensed band for low-powered communications.

Personal Digital Assistant   This is a hand-held or palm-top computer that contains intelligence sometimes known as an agent. It also often contains a wireless communications link in an unlicensed frequency for connecting the PDA to another computer on a network. An example of this is the Apple Newton II with a PDA software package.

Personal Information Processor   This is a scaled-down version of a personal communicator and typically does not include a voice communications capability. This is, in effect, the simplest form of personalized communications. It is sometimes referred to as an electronic tablet.

Personalized Communications   This is a computer with information processing capabilities that includes PDA capabilities and that is connected by wire or wireless links into a network.

Phased Array Antenna   This is a new type of communications antenna first developed by military communications research whereby very small antenna components designed for transmitting and receiving radio signals are networked together into a grid so that all of the components in the array work as a single antenna. This can allow the antenna to be flat and still perform as a directional and high-gain system by "electronic pointing and steering" of the electronic components whose performances are controlled by a digital processor.

Pioneer Preference   A special "head-start" award made by the FCC for innovators who develop unique new technologies. Motorola Iridium, which sought such status for their low earth orbit satellite system, did not receive it, but two pioneers in developing PCS service were granted this head-start award by the FCC in northern New Jersey and in southern California.

Push-to-Talk   This is a system whereby a single-frequency channel is used for both sending and receiving a mobile communications signal. This can also be used with different frequencies for transmit and receive for more advanced systems. The transfer from transmit to receive mode is activated by pushing a button on the hand-held communicator.

Radio Frequency (RF)   This is the part of the electromagnetic spectrum where electronic signals travel as radio waves. Within these frequencies communications can be achieved over the range from the low frequency band through the extremely high frequency band.

Roaming   This is the ability to travel beyond a local cellular mobile communications and to have your mobile telephone number forwarded to the locations where you are traveling.

Scrambling   The process of distortion of a radio carrier so as to make the signal "scrambled" so that unauthorized receivers cannot obtain a clear voice or video signal without having a descrambling device. This type of security system is not highly reliable and does not provide the quality of protection provided by full digital encryption. In video systems a transmission, in fact, looks very much like a scrambled image.

Short-Wave Frequency   This is also known as HF or High Frequency service (q.v.). This band is used for mobile communications and in some countries for telephone service. This band extends from 3 to 30 MHz and its wavelengths are long enough to be tolerant of variations in terrains and other obstructions in the transmission path, such as buildings, trees, and so on.

Signaling   The process for routing and interconnection of calls within a telecommunications system. At one time the signaling was provided within a telephone carrier and operated as an integral part of the overall calling process. This was called common-channel signaling. Advanced digital communications systems such as ISDN networks provide a separate signaling channel known as the D channel, which operates independently of the regular carrier channels known as B channels.

Signaling System Number 7   This is the CCITT/ISO-developed standard for advanced digital signaling that used the seven-layer protocol for internetwork connection.

Smart Card   A computer information storage card that can store a good deal of data for purposes of call routing, roaming, and billing. This type of card can be used with GSM, PCS, or other mobile services to facilitate mobile communications, especially when users are on travel.

SS/TDMA   Satellite or Space Switched Time Division Multiple Access techniques. This applies to satellite systems with on-board dynamic switching to interconnect different spot beams within the satellite within a TDMA channel assignment system. The INTELSAT VI and the Tracking and Data Relay Satellite System utilize this mode of operation.

Store and Forward   A form of data or message communications where a communication is transmitted to a computer memory and stored until it can be relayed to the intended final destination. In some communications satellite systems this can be several hours, while in conventional data networks the time in computer storage may be only a few milliseconds. This technique can be used as a tariff reduction scheme so as to relay fax or data messages during nonpeak, low-rate time periods.

Subscriber Identity Module (SIM)   This is a detachable unit from the subscriber mobile transceiver that is typically a "smart card" that allows easy data entry to allow roaming, remote billing, and many other "intelligent features." This is typically used within GSM mobile systems.

Super High Frequency   This is the frequency band from 3 to 30 GHz, which is heavily used for satellite communications and includes the C, X, Ku, Ka, K bands—all of which are used for space communications. This band is also referred to as the microwave band.

T-l Committee   This is the Committee of the American National Standards Institute of the United States that is charged with the development of telecommunications standards.

T1P1   The T-1 committee charged with developing mobile telecommunications standards.

TAC   A European-based mobile telephone standard that is now giving way to GSM standards.

Telecommunications Technical Standards Committee of Japan (TTC)   The standards-making unit for telecommunications in Japan that also has great influence throughout the Asia-Pacific region.

Telecommunications Industry Association   This is a cooperative industry group in the United States that promotes telecommunications trade and develops industry standards. Its TR-45 group is currently seeking to develop the standards for PCS services along with the ANSI-T1P1 group.

Telepoint   A low-power pedestrian-based "unlicensed" digital personal communications system used in the United Kingdom. The first version of Telepoint only allowed mobile reception of calls, but the second version allows for full interactivity. It has used "unlicensed or nonlicensed" frequencies for this service. The evolution of the various Telepoint systems are now known as CT-2, CT-2 Plus, and most recently, CT-3.

Temporary Mobile Subscriber Identity (TMSI)   This is a constantly updated identity number that substitutes for the IMSI and is used to establish location. The TMSI is constantly changed to ensure confidentiality of communications in a mobile environment.

Time Division Multiple Access (TDMA)   This is one of the most fundamental and well-proven digital multiplexing techniques. Time division modulation does not depend upon changing frequencies or amplitude to create the signal. Rather, it involves sharing the frequency in the time domain and assigning channel assignments to users in very precisely defined time slots measured in milliseconds.

Transceiver   A transceiver is a communications device that is capable of transmitting a signal as well as receiving a signal from another source. This can be as simple as a walkie-talkie device up to a much more sophisticated communications unit with a relatively large antenna system.

Transmission Control Protocol/INTERNET Protocol (TCP/IP)   The protocol for the INTERNET system that is a simple, low-overhead transport system optimized for UNIX-based computer networking.

Transponder   The active communications portion in a radio relay system, such as a satellite. The transponder receives the incoming signal, filters it, translates the incoming frequency to the outgoing frequency, amplifies it, and sends the signal forward to the antenna system for relay to the intended destination.

TT Committee of Japan   The national telecommunications technical standards committee of Japan.

Ultra High Frequency   This is the frequency band from 300 to 3000 MHz. This frequency is used for television distribution, mobile communications, and a variety of other applications. Because of the physical characteristics of the waves in this band and their communications carrying capabilities, UHF (as well as VHF) are particularly well suited to mobile communications.

Ultra Small Aperture Terminal   A satellite earth station antenna usually for Fixed Satellite Service that is anywhere from 25 to 100 cm in size. This is often used for remote data collection.

Universal Mobile Telecommunications Service (UMTS)   The move to create a global standard for personal communications that would consolidate existing competing standards. This initiative to promote UMTS has been sponsored by the European Telecommunications Standards Institute (ETSI) and is closely aligned with the longer-term standards initiative of the ITU known as Future Planned Land Mobile Telecommunications Service (FPLMTS). The concept of UMTS is to create a device that can be used at home, at work, or in any public place that would combine phone, messaging, paging, and most enhanced services.

Universal Personal Telecommunications (UPT)   The concept of allowing everyone to have their own personal communicator and their own individual telephone number. This would combine into an integrated service all technologies such as cellular, PCS, satellites, and fixed terrestrial services. It is not clear whether this will be accomplished through a decentralized telecommunications systems model or a centralized model. (See below.)

Universal Telephone Number   The world of telecommunications is becoming more and more global. Further, as the system evolves from today's environment of calling a location rather than a person, the concept of a universal telephone number will become more important. In the new UTN environment each individual will have their own 11-digit telephone number that ultimately will be valid no matter where they may be in the world.

Unlicensed PCS   There will be a 40-KHz band between 1890 and 1930 MHz reserved for unlicensed Personal Communications Services. This will be segmented into a voice applications band and a data transmission band. Unlike the exclusive-use licensed band, these frequencies will be available for use by anyone who abides by the rules of use as established by the FCC.

Up-link   The portion of a complete satellite link that connects the transmitting ground segment antenna to the satellite.

Very High Frequency   This is the frequency band that spans the range of 30 to 300 MHz. It is used in the United States primarily for over-the-air television distribution and some mobile services. In other parts of the world, especially developing countries, the HF and VHF bands are used for voice communications.

Very Low Frequency   This is the lowest usable frequency band, which ranges from 3 Hz to 30 KHz.

Very-Small-Aperture Terminal (VSAT)   A Fixed Satellite Service ground antenna that is typically between 1 and 3 meters in size and that is typically used in a large corporate communications network.

Visitor Location Register (VLR)   This is a database that is constantly updated with key information on subscribers who are roaming to mobile service areas outside of their home system.

Wireless Local Area Network (LAN)   This is a radio-based or wireless service that is used to connect computers, digital telephones, and other digital devices together in a corporate office environment. The connectivity for the LAN rather than being a wire or fiber link is provided by a radio transmitter/receiver bus, usually in an unlicensed frequency band and subject to maximum transmitter power levels.

Wireless Public Automatic Branch Exchange (PABX)   This is a radio-based or wireless service that is used in corporate office environment to take the place of a conventional wire-based PABX system. The equipment and functionality of Wireless LANs and PABXs are, in fact, very close, especially within an all digital office.

World Administrative Radio Conference   These are periodic plenipotentiary conference meetings of all members of the ITU with treaty-making powers. The primary purpose of these conferences is to reallocate the use of radio frequencies for worldwide and regional communications purposes. There are regional versions of these conferences for ITU regions 1, 2, and 3 and these are known as Regional Administrative Radio Conferences.

APPENDIX THREE

STANDARDS MAKING FOR PERSONAL COMMUNICATIONS SERVICES (PCS)

- Voice
- Medium-rate data (< 144 kb/s)
- Pedestrian/vehicular
- Indoor/outdoor
- Urban, suburban, rural

FIGURE A.1 PCS Services

PCS Standards
- ANSI T1 and Associated Technical Subcommittees
- Telecommunications Industries Association (TIA)
- ITU-S
- ITU-R

Other
- Personal Communications Industries
  Association (PCIA)
- Cellular Telecommunications Industries Association (CTIA)
- WIN FORUM
- IEEE 802.11

FIGURE A.2 Standards Organizations Involved in PCS

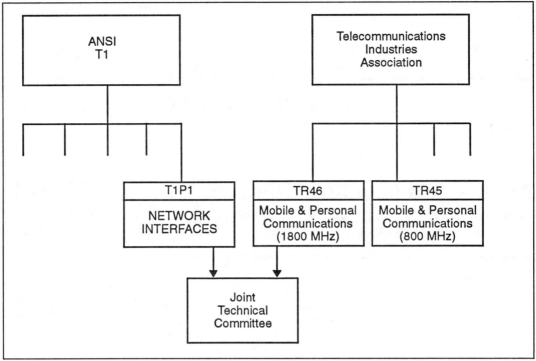

FIGURE A.3   Common Air Interface Standards Development

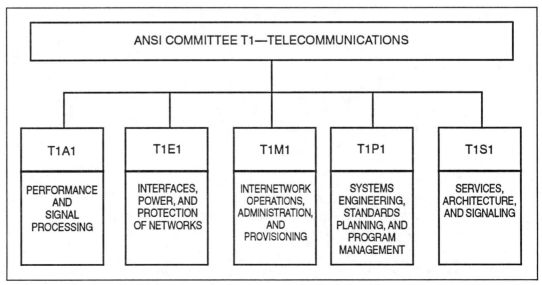

FIGURE A.4   The ANSI Telecommunications Standards Committee Structure

- Frequency Band
- Access Method (CDMA, TDMA)
- Duplex Method
- Modulation Technique
- Portable Handset Transmit Power
- RF Channel Spacing
- Diversity Techniques
- Handover Technique

FIGURE A.5   Major Common Air Interface
Technical Parameters

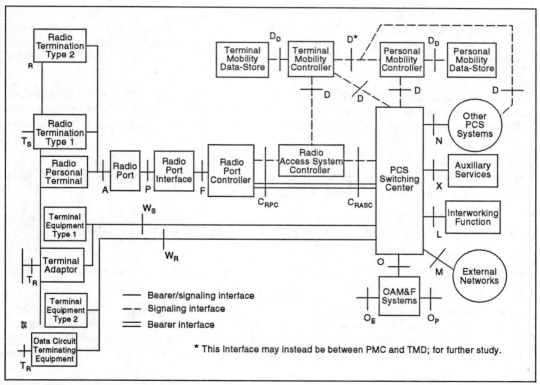

FIGURE A.6   PCS Reference Architecture

5.2   Reference Points

5.2.1   *A* Reference Point

Radio connection between an RP and a Radio Personal Terminal that carries user voice and data along with call, connection, service, system, and local (RF) control signaling.

5.2.2   *C* Reference Point

Combination of Reference Point C_B and Reference Point C_{S1} reference points.

5.2.2.1 C_B Reference Point

Access facilities between a PSC and an RPC. It is expected that different variants of Reference Point C_B will provide differing degrees of service support.

5.2.2.2 C_{S1} Reference Point

Control signaling between a RASC and a subtending RPC. This includes local control signaling between the RASC and the RPC as well as common control signaling between the RASC and a radio personal terminal.

5.2.3 C_{S2} Reference Point

Control signaling between an RASC and a PSC to support basic call and connection control, basic and supplementary services, and the transfer of terminal and user identification associated with Reference Point C_B access facilities.

5.2.4   *D* Reference Point

Signaling between an RASC, PSC, or PMC and a TMC, or a PSC and a PMC for the exchange of radio terminal-related or user-related information to support authentication, service validation and delivery, location management, call management, routing, and so on. This signaling may also be used between PCS systems.

5.2.5 D_D Reference Point

Data exchange signaling between a TMC and a TMD, or a PMC and a PMD for the exchange of radio terminal-related or user-related information.

5.2.6   *F* Reference Point

Facilities and associated control signaling between an RPC and a subtending RPL.

5.2.7   *L* Reference Point

Set of interfaces between a PSC and External Networks supporting data interworking.

5.2.8   *M* Reference Point

Interswitch facilities and associated control signaling between PSCs to support basic call, service, and connection control, and the transfer of terminal and user information and possibly additional information related to these. This interface supports, for example, handover between PSCs or mobility processing by a remote PSC. This distinguishes the Reference Point *M* from the Reference Point *N*.

FIGURE A.7   PSC Reference Points

5.2.9 *N* Reference Point

Interswitch facilities and associated control signaling to External Networks to support basic call, service, and connection control.

5.2.10 *O* Reference Point

Set of interfaces carrying operational information to support monitoring, testing, administration, traffic, and billing requirements.

5.2.11 O_P Reference Point

Set of secure interfaces, partitioned by PCS Service Provider and capability (e.g., display, edit, add/delete), carrying operational information to support a PCS Service Provider's requirements for network management, testing, administration, traffic, and billing.

5.2.12 O_E Reference Point

OAM&P interface to each of the reference elements in the reference model.

5.2.13 *P* Reference Point

Facilities and associated control signaling between an RPI and a subtending RP.

5.2.14 T_R Reference Point

Non-ISDN terminal equipment interface.

5.2.15 T_S Reference Point

ISDN terminal equipment interface.

Note: The "*T*" reference point is not the ISDN "*T*" interface.

5.2.16 W_R Reference Point

Non-ISDN wireline access interface.

5.2.17 W_S Reference Point

ISDN wireline access interface.

5.2.18 *X* Reference Point

Facilities and/or control signaling between a PSC and any of a variety of auxiliary services such as voice mail, paging, and so on.

FIGURE A.7 (CONT.)   PSC Reference Points

ABOUT THE AUTHOR

JOSEPH N. PELTON

Dr. Joseph N. Pelton is the Director of the oldest and largest graduate telecommunications program in the United States, with over 400 students now enrolled in the Interdisciplinary Telecommunications Program at the University of Colorado at Boulder. He is also Director of the Center for Advanced Research in Telecommunications (CART), which is the research arm of the Telecommunications Program with ongoing research being carried out for the U.S. government (NASA and NTIA), Network World, Northern Telecom, Inc., the State of Colorado, and many other organizations. Dr. Pelton spent almost 22 years as an executive and manager for the COMSAT Corporation and the International Telecommunications Satellite Organization (INTELSAT).

Dr. Pelton is the author of 12 books in telecommunications, satellite communications and the future. His books have been nominated for several awards, including a Pulitzer Prize nomination. *Global Talk* won the American Astronautics Society's Literature Prize of 1984. He received the B.S. degree in Physics from the University of Tulsa, the M.A. degree in International Relations from New York University, and the Ph.D. degree in Political Science from Georgetown University.

Dr. Pelton currently serves as a Full Professor with the International Space University (ISU) headquartered in Strasbourg, France, and also serves as Chairman of the Board of Trustees of this international educational institution. He has also been elected a member of the International Academy of Astronautics, Who's Who International, and the International Biographical Institute. He is a former trustee of the International Institute of Communications and the Managing Director of the Arthur C. Clarke Foundation of the United States.

In 1983 he was appointed Managing Director of the U.S. Committee for World Communications Year by President Reagan. In 1984 he founded the Society of Satellite Professionals and acted as its first President. The Society awards an annual Scholarship in his honor. In 1987 he was awarded the H. Rex Lee award of the Public Service Satellite Consortium for organizing and directing INTELSAT's Project Share activities, which brought tele-health and tele-education projects to 63 countries and gave rise to the Chinese National Television University with its three million students.

INDEX